Food Security and
Farm Land Protection
in China

Series on Chinese Economics Research

(ISSN: 2251-1644)

Series Editors: Yang Mu *(National University of Singapore, Singapore)*
Fan Gang *(Peking University, China)*

Published:

Vol. 1:   China's State-Owned Enterprises: Nature, Performance and Reform
by Sheng Hong and Zhao Nong

Vol. 2:   Food Security and Farm Land Protection in China
by Mao Yushi, Zhao Nong and Yang Xiaojing

Series on Chinese Economics Research – Vol. 2

Food Security and Farm Land Protection in China

Mao Yushi
The Unirule Institute of Economics, China

Zhao Nong
The Unirule Institute of Economics, China
Chinese Academy of Social Sciences, China

Yang Xiaojing
The Unirule Institute of Economics, China

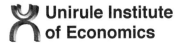

Unirule Institute
of Economics

World Scientific

Published by

World Scientific Publishing Co. Pte. Ltd.

5 Toh Tuck Link, Singapore 596224

USA office: 27 Warren Street, Suite 401-402, Hackensack, NJ 07601

UK office: 57 Shelton Street, Covent Garden, London WC2H 9HE

Library of Congress Cataloging-in-Publication Data
Mao, Yushi.
 Food security and farm land protection in China / by Mao Yushi, Zhao Nong, Yang Xiaojing.
 p. cm. -- (Series on Chinese economics research ; v. 2)
 ISBN 978-9814412056
 1. Agriculture--Economic aspects--China. 2. Land use, Rural--China.
3. Agriculture and state--China. 4. Food supply--China. 5. Food supply--Government policy--
China. I. Zhao, Nong. II. Yang, Xiaojing. III. Title.
 HD2098.M355 2013
 338.1'951--dc23

                                                                  2012036373

British Library Cataloguing-in-Publication Data
A catalogue record for this book is available from the British Library.

In-house Editor: DONG Lixi

Typeset by Stallion Press
Email: enquiries@stallionpress.com

Printed in Singapore.

About the Authors

MAO Yushi is the director-chairman of board of the Unirule Institute of Economics, an independent think tank in China, which he and four other economists founded in 1993. His previous appointments were with the Railway Research Institute of China, and the Chinese Academy of Social Sciences (CASS). He was a visiting senior lecturer at University of Queensland, Australia and a visiting scholar at Harvard University. From 1987 to 1994, he was appointed as a resource person for the Africa Energy Policy Research Network. In 2004, he was elected by the International Business Review as one of the ten most influential economists in China. He received the Cato Institute's Milton Friedman Prize in 2012 for Enhancing Liberty for his works in classical liberalism and free-market economics. His research areas are institutional economics, energy and environmental economics, transportation, policies on economic reform and poverty alleviation. His publications are mostly in Chinese including: *Economics in Everyday Life* (1993); *The Future of Chinese Ethics* (1997); *The Principle of Optimal Allocation: Mathematical Foundation of Economics* (1985); *Who Bars Us from Getting Rich* (1994); *Searching for the Road to Increase Social Welfare* (2002); *Economics in Plain Language* (2003); *Give Freedom to Those Who You Love* (2003); *The Economics I Understand* (2004); *Ethics, Economics, and Institution* (2002); *Ten Lectures in Micro-economics* (2004); *The Wisdom in Economics* (2004); *Understanding the World around Us* (2007).

ZHAO Nong is a research fellow of the Institute of Economics, CASS and the Unirule Institute of Economics, and a member of the Academic Committee of the Unirule Institute of Economics. His research interests are institutional economics and industrial organization theory. His publications include *Barrier for Entry and Exit: Theory and Application* (2007) and *What Can Protect China's Agricultural Security* (2011).

YANG Xiaojing, is a PhD candidate in Economics, Renmin University of China. Her research interests are institutional economics, public policy and agricultural economics.

Series on Chinese Economics Research Editorial Committee

Contents

List of Figures

List of Tables

Preface

People's thinking always lags behind reality. Particularly, in this present era of rapid changes policies are often based on past experiences and are therefore unable to take into account today's circumstances. One such example is food security.

The most devastating famine in human history occurred in China in the late 1950s. That is why our present political rulers remain constantly alert to the danger of famine. The current situation is, however, completely different from that of 50 years ago, and the conditions and environment both at home and abroad have completely changed. Domestically, the problem of food production and distribution has already been fully solved. Internationally, there is an ample supply of food in the globalized market. The possibility of a recurrence of famine in China is slim, if not zero. China's current food policies are, however, still based on the assumption that a famine may occur anytime, anywhere.

Most worrying is the fact that certain people associate food security with farmland protection and insist that the red line of 1.8 billion mu of cultivated land shall never be broken. After 30 years of reforms, China's population has grown by 45% and grain output increased by 60%, while the total area of cultivated land has decreased. It is clear that there is no direct relationship between farmland area and grain output. An immediate consequence of farmland protection policy is the dramatic rise in real estate prices. In other words, increase in real estate prices is not due to expensive reinforced concrete or high labor costs, but because of limited land supply. Another long-term consequence is the delay of China's urbanization process. After all, it is impossible for hundreds of millions of farmers to move into cities without any farmland being occupied.

The association of food security with farmland area is an invention of the American scholar Lester Brown. According to his confusing logic, the construction of too many golf courses in China could lead to a shortage of food production. What he does not know is that basketball courts in universities and middle schools were all planted with food crops during the Great Famine in China. It is difficult to relate the proliferation of golf courses with overproduction or shortage of food. Certainly, we cannot lay all the blame on foreigners for China's misguided policy. The land administrative departments of our country favor land control mainly because of huge departmental interests.

It is an obligation of a private think tank to offer alternate viewpoints on such a major national policy. That is the reason why we have applied for a grant from the Ford Foundation. Initially, we planned for a rather small-scale research project and did not expect that it would involve so many aspects. Eventually, it evolved into a huge project in which several topics were included, such as the lessons from the famines in China and that of other countries in the world, the reliability of the international food supply, consequences of historical food embargos, an analysis of factors determining food output, balance in world food production and demand, historical fluctuations of food prices, etc. This unexpected achievement presents a thorough analysis of China's food situation and provides extremely useful basic data for formulating national policies concerning food and farmland.

Trustworthy conclusions can only be reached through tedious verification of facts in a serious research work. We hope that the research results presented in this book will be helpful in clarifying the debate on this topic.

Mao Yushi
Unirule Institute of Economics

Chapter 1

Research on the Total Area, Structure and Quality of China's Cultivated Land

1. Introduction

Cultivated land refers to land that is exploited and regularly cultivated for planting agricultural crops. Strictly speaking, cultivated land refers to land used for planting agricultural crops, including regularly cultivated land, newly exploited land, reclaimed and consolidated land, fallow land (including shifting fallow land and shifting cultivated land). Cultivated land is mainly used for planting crops (including vegetables), scattered with fruit trees, mulberry trees and other trees; cultivated land also includes (a) cultivated sea-beaches with at least one harvest per year; (b) ditches, canals, paths and field banks with the width of less than 1.0 meter in the south of China, and 2.0 meters in the north of China. Cultivated land is temporarily used for planting herbs, turfs, flowers, nursery stocks and so on, as well as used temporarily for other purposes.

The quality of cultivated land is better than that of forest land and grassland because some strict conditions have to be met for a piece of land to be qualified as cultivated land, such as topography, soil, moisture content, temperature, etc. That is why cultivated land possesses the widest adaptability and highest productivity. Although China's cultivated land does not take up a high proportion of the total national agricultural land area (18.5%),[1] its productivity accounts for approximately three-fourths of the productivity of the national agricultural land. This report summarizes

---

[1]This proportion is calculated based on the results of the Land-use Change Survey made by the Ministry of Land and Resources of China in 2006, in which the total area of cultivated land was reported to be 1.827 billion mu, while the total area of agricultural land was 9.558 billion mu.

the current situation of China's cultivated land, and makes an analysis from the perspectives of the total area, structure and quality changes of cultivated land.

The structure of this chapter is as follows: The first part is an introduction which presents the definition of cultivated land; the second part depicts the current situation of China's cultivated land; the third part, which is the focus of this report, gives a detailed description of the changes of China's cultivated land area since its establishment; the fourth part depicts the structural changes of China's cultivated land; followed by an analysis from the perspective of cultivated land quality, pointing out that total area, structure and quality are three major aspects of this research on cultivated land; the last part makes an analysis of the project of converting cultivated land to forests in recent years.

2. The Current Situation of China's Cultivated Land

2.1. *China's land and agricultural land: Categories and total area*

According to the results of the Land-use Change Survey conducted by the Ministry of Land and Resources of China in 2006, there are 9.858 billion mu (1 hectare = 15 mu) of agricultural land (accounting for 69.1% of the total), 485 million mu of construction land (3.4%) and 3.917 billion mu of unused land (27.5%) (see Fig. 1.1).

Land Classification and areas (0.1 billion mu)

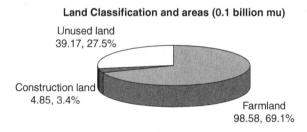

Unused land
39.17, 27.5%

Construction land
4.85, 3.4%

Farmland
98.58, 69.1%

Figure 1.1. China's Land Structure.

Source: Land-use Change Survey by the Ministry of Land and Resources of China in 2006, adapted by the author.

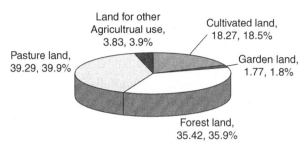

Land Classification and areas (0.1 billion mu)

Land for other Agricultrual use, 3.83, 3.9%

Cultivated land, 18.27, 18.5%

Pasture land, 39.29, 39.9%

Garden land, 1.77, 1.8%

Forest land, 35.42, 35.9%

Figure 1.2. China's Farmland Structure.
Source: The Investigation Results of Land-use Change Survey Carried Out by the Ministry of Land and Resources of China in 2006, adapted by the author.

Specifically speaking, among the agricultural land, there are 121.7759 million hectares (1.827 billion mu) of cultivated land, 11.8182 million hectares (177 million mu) of garden land, 236.1213 million hectares (3.542 billion mu) of forest land, 261.9320 million hectares of pasture land, and 25.5410 million hectares (383 million mu) of land used for other agricultural purposes (see Fig. 1.2).

2.2. Distribution of China's cultivated land

About 88% of the China's total cultivated land is distributed in the moist and semi-moist areas east of the Daxing'an Mountain Range–Zhangjiakou–Yulin–Lanzhou–Changdu line. According to the statistics of agricultural land classification, the proportions of cultivated land in descending order are as follows: Huang He–Huai He–Yellow Sea Region and the Middle-Lower Yangtze River Region take up the highest proportion, followed by Northeast region and Southwest region; the Loess Plateau, Inner Mongolia and Great Wall Vicinity, South China Region, Ganxin Region and Qinghai–Tibet Plateau Region, take up the lowest proportions.

Cultivated land can also be divided into paddy fields and dry farmland according to its irrigation conditions and utilization. The area of paddy fields accounts for 25.34% of the total cultivated land area. Over 90% of the total paddy field area is located South of the Qinling Mountain Range–Huaihe River line, mainly in the Middle-Lower Yangtze River Plain, the Sichuan Basin, the Pearl River Delta, and Chaoshan Plain. 22.67% of the dry farmland are irrigated lands with high and stable yields. Irrigated lands are mainly distributed in North China Plain, Fen He–Wei He Plains, Hetao Area, the West Huang He Corridor and the agricultural areas at the feet of the Tianshan Mountains in the Xinjiang Uygur Autonomous Region. The rest of the dry farmland is rain-fed dry-cultivated land in terrains where irrigation is impossible and relies solely on rainfalls. Therefore, yields of this sort of dry farmland are low and unstable. Rain-fed dry farmland is located all over the country and mainly in Northeast Plains, the Loess Plateau and the transition zones between cropping areas and nomadic areas.

Cultivated land can also be divided into flat cultivated land and sloping cultivated land according to the gradient of its terrain. Flat cultivated land is flat and highly productive, and is mainly located in the plains of eastern China, with the largest areas in the Huang He–Huai He–Yellow Sea Plain, the Middle-Lower Yangtze River Plain and the Northeast Plain. The maximum slope is 25°, beyond which the terrain will be unsuitable for farming. Sloping cultivated land is mainly distributed in mountainous and hilly areas, with the largest area in Southeast district and the second largest areas in the Loess Plateau and the Middle-Lower Yangtze River hilly areas.

2.3. Total area and classification of China's cultivated land

The *Communiqué on Major Statistical Data of the Second National Agricultural Census of China*, jointly issued by the Office of the Leading Group of the State Council for Second National Agricultural Census, Ministry of Land and Resources, and National Bureau of Statistics on February 29, 2008 discloses the latest statistics on the total area, distribution and classification of China's cultivated land. For details, please see Table 1.1.

In 2006 (as of October 31, 2006), China's cultivated land area (excluding Hong Kong Special Administration Region, Macao Special Administration Regions and Taiwan province) amounted to 121.7759 million hectares. In terms of distribution by region, cultivated land distributed in the Western region takes up a high proportion of 36.9%; that distributed in Eastern, Central and Northeastern regions account for 21.7, 23.8 and 17.6% respectively. In terms of classification of cultivated land, dry farmland occupies 55.1% of China's total cultivated land; paddy fields occupy 26.0%

Table 1.1. Statistics on Distribution and Classification of China's Cultivated Land.

	Areas (1,000 hectares)	Proportion (%)
Total	121,775.9	100.0
By regions		
Eastern region	26,395.2	21.7
Central region	28,991.6	23.8
Western region	44,937.9	36.9
Northeastern region	21,451.2	17.6
By classifications		
Paddy fields	31,667.9	26.0
Irrigated farmland	22,963.3	18.9
Dry farmland	67,144.7	55.1
By gradients		
0–15°	106,591.8	87.5
15–25°	11,143.2	9.2
of which: terraced	3,177.5	—
over 25°	4,040.9	3.3
of which: terraced	900.3	—

Note: Details of the four regions: Eastern Region includes Beijing, Tianjin, Hebei, Shanghai, Jiangsu, Zhejiang, Fujian, Shandong, Guangdong, Hainan; Central Region covers Shanxi, Anhui, Jiangxi, Henan, Hubei, Hunan; Western Region includes Inner Mongolia, Guangxi, Chongqing, Sichuan, Guizhou, Yunnan, Tibet, Shaanxi, Gansu, Qinghai, Ningxia, Xinjiang; Northeastern Region covers Liaoning, Jilin, Heilongjiang. *Source*: *Communiqué on Major Statistical Data of the Second National Agricultural Census of China.*

and irrigated farmland occupies 18.9%. In terms of gradient, cultivated land between 0° and 15° makes up 87.5% of China's total cultivated land area; that between 15 and 25 and over 25° occupies 9.2% and 3.3% respectively.

3. Changes in China's Cultivated Land Area since its Establishment

3.1. Changes in China's cultivated land area since its establishment: Domestic statistics

In terms of time series data of the total area of China's cultivated land, *China Statistical Yearbook* provides the longest time series data (1949–present). However, land area data provided by *China Statistical Yearbooks* were lower than the actual data, as was noted in the Yearbook. The changing trend in

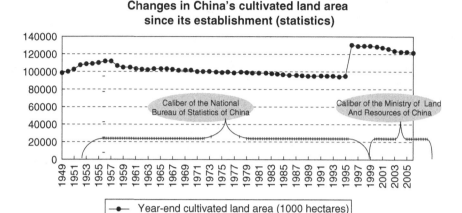

Figure 1.3. Changes in China's Cultivated Land Area Since Its Establishment. *Source: Compiled by the author according to statistical data of China Statistical Yearbook and China Agricultural Development Report 2007.* For details, please see attached the Appendix.

China's cultivated land area in the past 60 years since the founding of the People's Republic of China, which is based on statistics of the year-end total cultivated land area reported in *China Statistical Yearbook* as well as relevant information in *China Agricultural Development Report 2007*, is shown in Fig. 1.3.

It should be noted that the traditional data of China's cultivated land area is dramatically smaller than the satellite remote sensing data. From Fig. 1.3, we can see that China's cultivated land area in the year 1995 was 94.9709 million hectares according to *China Statistical Yearbook 1996*, while that estimated by the Ministry of Land and Resources of China in October 1996 according to satellite remote sensing data was 130.0392 million hectares, 35.0683 million hectares or 36.9% higher than the released data of 1995. Data after 1996 were based on the detailed land survey released by the Ministry of Land and Resources of China. Although data of cultivated land area released by the National Bureau of Statistics of China was smaller than actual data, the National Bureau of Statistics of China possesses yearly provincial level data since the founding of new China and there is uniformity in its statistical figures before 1995. Therefore, the changing trend of China's cultivated land area as reflected by the data of the National Bureau of Statistics of China is basically reliable.[2] On the other hand, due to its accuracy and uniformity over time, data of the cultivated land area released by the Ministry of Land and Resources since 1996 are gradually being accepted by people from all walks of life.

Due to the change in statistical sources, we divide our analysis into two phases, the phase prior to 1996 and the phase after 1996.

According to the statistics of the National Bureau of Statistics, China's cultivated land area was 97.88133 million hectares in 1949; it peaked at 111.83 million hectares in 1957. Since then it experienced a continual decline and reached 99.38933 million hectares in 1978; thereafter, it descended at a slower rate to reach 98.35933 million hectares by the end of 1983 with an average annual reduction of 206,000 hectares. The decline rate started to rise again after 1984, and the total area shrunk to a mere 94.9739 million hectares by the end of 1995 with an average annual decrease of 280,000 hectares. Most notably, 1.6 million hectares of cultivated land was occupied in 1985, resulting in a net loss of cultivated land of 1.01 million hectares, the largest net loss since the founding of the New China. In addition, about 16.9 million hectares of cultivated land was lost all over the country from 1955 to 1957.

In general, China's cultivated land area shows an evident reduction trend after 1996, resulting in another slide in China's cultivated land resources. The total area dropped by 6.35% from 130.0392 million hectares in 1996 to 121.7759 hectares in 2006 with an accumulated loss of 8.2633 million hectares.

3.2. Changes in China's cultivated land area since its establishment: International statistics from FAO

It can be seen from Figs. 1.4 and 1.5 that data of *China Statistical Yearbook* is similar to the FAO data before 1982, and the turning point is the year 1983,

---

[2]Liang Shumin (2006). The cultivated land resources under the background of urbanization. *Economic Reference*, 21 January issue. Available at http://202.84.17.25/www/Article/20061218154-1.shtml.

Changes in China's cultivated land area in accordance with statistics of FAO

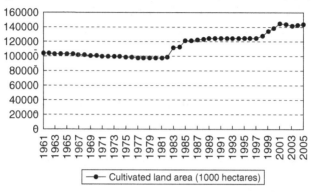

Figure 1.4. Changes in China's Cultivated Land Area, 1961–2005.
Source: FAOSTAT, compiled by the author. For details, please see Appendix 5.

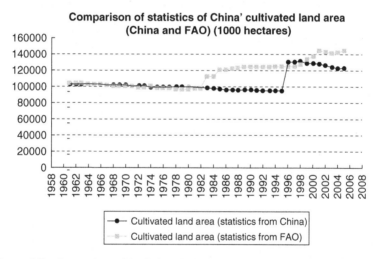

Figure 1.5. Comparison of Statistics of China's Cultivated Land Area, 1961–2005.
Source: Compiled by the Author.

when China's cultivated land area was 98.35933 million hectares according to data of *China Statistical Yearbook*, and 111.333 million hectares according to data of FAO, with a significant difference of 12.86167 million hectares. As the statistical sources of FAO are still unknown, this huge difference cannot be explained.

3.3. Causes of changes in cultivated land area and reliablity analysis of relevant data

Statistics of China's cultivated land area according to *China Statistics Yearbooks* provide a rather accurate reflection of the trend of changes, although the land area figures are smaller than actual data as mentioned above. In the following paragraphs, we will analyze the changing trend of China's cultivated land area at different phases after the founding of New China. We will also assess the reliability of the statistics.

3.3.1. *1949–1957: Significant increase of cultivated land area*

In the first few years after the founding of New China, China's cultivated land area increased substantially for the following three main reasons. First, due to the nationwide land reform between 1949 and 1952 and the encouragement of the policy of "anyone who reclaims the land will own the harvest of the land", the farmers' enthusiasm for production was greatly aroused and a large amount of cultivated land abandoned due to wars was reclaimed. Second, Mao Zedong issued a decree demanding the army to participate in production and construction work in December 1949, leading to the increase of the total area of reclaimed land. Third, from 1953 to 1957, the state set up many state farms and army reclamation farms, encouraged young intellectuals to migrate to rural areas to participate in voluntary reclamation of wasteland, and called on farmers to participate in reclamation of wasteland in local regions. As a result, 5.4765 million hectares of land was reclaimed in total during these four years. After deduction of land occupation for urban and rural construction and land deserted due to disasters, China's cultivated land area rose to 111.83 million hectares in 1957 (Zhang Shigong and Wang Ligang, 2005).

Following is an analysis of the reliability of the statistical data. The statistics of China's cultivated land area in 1949 was derived by making use of the statistics of the Kuomintang government plus an estimation of the time, its reliability is rather poor. During the land reform between 1949 and 1952, cultivated land areas were remeasured, however, the raw data at the grassroot level have not been gathered together. From 1951 to 1953, the program of surveying land and fixing production quota was carried out all over the country in order to collate and stipulate agricultural tax, and China's total cultivated land area was calculated by summing up the results of this survey. This figure was accurately reflected in the statistics for the year 1953, i.e., 108.53 million hectares. The statistics for the year 1953 best represented the real total area of cultivated land in China in the early years after liberation. The total area of cultivated land was substantially increased through large-scale reclamation of wasteland during 1953 and 1957, and the trend of changes reflected by statistics is rather credible (Feng Zhiming *et al.*, 2005; Fu Chao *et al.*, 2007).

3.3.2. *1958–1960: Drastic decrease of cultivated land area*

The period from 1958 to 1960 is an extremely abnormal period during which the total area of cultivated land witnessed a drastic decrease. For one thing, the Great Leap Forward resulted in disproportions in the national economy; vast area of cultivated land was either occupied or deserted. On top of this, the Three Years of Natural Disasters worsened the situation. The total area of cultivated land dropped from the peak of 111.83 million hectares in 1957 to 104.86133 million hectares in 1960.

After making a comparison with the estimated data issued in 1978 by the Ministry of Agriculture and Forestry, Feng Zhiming *et al.* (2005) believed that the net change of cultivated land area reflected by *China Statistical Yearbook* was basically reliable.

3.3.3. *1960–1978: Reliability of statistics comes under question*

Statistics show that the total area of cultivated land descended continually after 1960 and fell to 99.30533 million hectares by 1980.

Some scholars doubted the reliability of the statistics of cultivated land area of this period, and many scholars believe that there is a big difference between the statistics and the actual cultivated land area from 1960 to 1978. Therefore, the reliability of the continual decrease in the cultivated land area between 1960 and 1980 shown by the statistics is questionable.

According to various land use surveys carried out separately by many departments and organizations in the early 1980s, China's cultivated land area in the early 1980s was 132.5 to 139.7 million hectares (please refer to Table 1.2).

Bi Yuyun and Zheng Zhenyuan made an analysis of the increase and decrease of China's cultivated land area since 1949. In their research, the increase of cultivated land area between 1953 and 1980 was estimated on the basis of relevant reference materials. The decrease of cultivated land area was classified as cultivated land occupied by construction, cultivated land occupied by agricultural land structural adjustment, cultivated land destroyed by disaster and deserted cultivated land. By summing up the various categories listed above, they estimated the area of cultivated land occupied for other purposes. The result of their research showed that the newly increased cultivated land area prior to 1980 exceeded 22.89 million hectares while the reduced area stood between 17 million and 21 million hectares, with the increase exceeding the reduction.

The reasons for the distortion of statistics of cultivated land area during this period are as follows. First, since vast reduction of cultivated land area would lead to reduction of agricultural tax income, a regulation was issued stipulating that decrease of cultivated land area should only be reported when exemption of agricultural tax was involved, whereas occupied cultivated land not qualified for agricultural tax exemption should not be reported. Second, at the time concerned, in order to encourage wasteland reclamation, the government permitted newly reclaimed land to be excluded from the area of cultivated land within a specific grace period of time. However, most of the reclaimed wasteland was not reported even after the expiration of the grace period. Third, in order to inflate the per unit area yield, large amounts of cultivated land were used as "helping farmland", their yields were reported while their areas were not reported. Fourth, areas of disaster destroyed cultivated land were exaggerated in order to gain more aids from the state. Fifth, although large amounts of cultivated land requisitioned for state construction was deducted from the statistics, some of these requisitioned land continued to be cultivated because a large number of these construction projects did not materialize. Hence, the discrepancy between the statistics of China's cultivated land area and the actual land area became larger and larger, resulting in the false statistical impression that China's cultivated land area had decreased year by year.

Table 1.2. Statistical Survey of China's Cultivated Land Area Since 1980 (10,000 Hectares).

Data source	Agricultural regionalization committee of China	1:1000000 map of Chinese land resource	National resources & environment survey using remote sensing	The second national soil survey of China	The first national land survey
Host unit	Agricultural Regionalization Committee of China	Commission for Integrated Survey of Natural Resources of China Academy of Sciences	Institute of Remote Sensing Applications of China Academy of Sciences	National Soil and Fertilizer Station of the Ministry of Agriculture of China	China Land Administration Bureau
Relevant period	Early 1980s	Early 1980s	Early 1980s	1985	1996
Cultivated land area	13,969	13,906	13,782	13,252	13,003.9

Source: Feng Zhiming *et al.* (2005).

3.3.4. *1979–1999: The continual decrease of cultivated land area shown by statistics is questionable*

The data of the increase and decrease of cultivated land area released by the Ministry of Land and Resources of China (named the State Bureau of Land Administration before 1997) ever since 1987 basically reflected the true tendency of changes of China's cultivated land resources.

As mentioned above, there was a statistical source change in 1996. According to the statistics of *China Statistics Yearbook*, the total area of cultivated land declined from 99.498 million hectares in 1979 to 94.0739 million hectares in 1995, while according to the statistics of the Ministry of Land and Resources, the total area of cultivated land decreased from 130.0392 million hectares in 1996 to 121.8 million hectares in 2006 (see Fig. 1.6).

China's cultivated land area showed an obvious decreasing trend since 1996 and this trend was not put under control until after 2004. The substantial reduction in cultivated land area was mainly caused by conversion of cultivated land to forests, which will be analyzed separately in the last part of this chapter.

Deng Xiangzheng *et al.* (2005) used the Land Use Types (LUT) dataset of the Chinese Academy of Sciences and analyzed the satellite remote sensing data between 1986 and 2000. According to their research, during this period 3.06 million hectares of cultivated land were converted to other types of land while over 5.7 million hectares of other types of land were converted to cultivated land. As a result, the total area of cultivated land actually increased from 1986 to 2000.

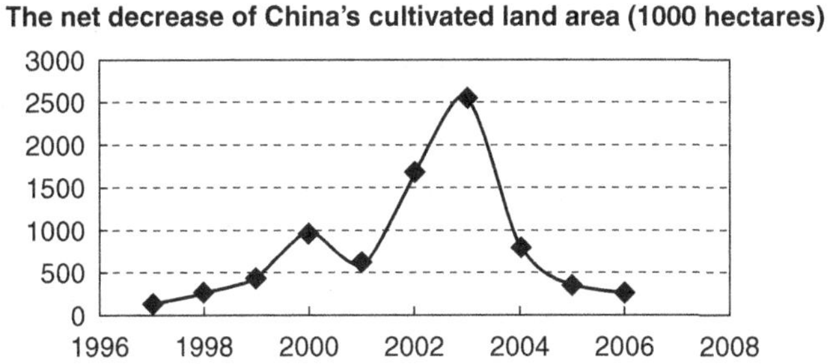

The net decrease of China's cultivated land area (1000 hectares)

Figure 1.6.   The Net Decrease of China's Cultivated Land Area, 1996–2006.

3.4. *Partial correction to the total area of cultivated land*

In terms of making correction to China's cultivated land area, data of three periods is credible and can be used as the basis for data correction. First, China's cultivated land area for the year 1953, calculated through summarizing the statistical data resulted from the land survey to fix production quota, was accurately reflected in the statistics for that year, i.e., 108.53 million hectares. Second, according to the land use surveys carried out separately by many departments and organizations in the early 1980s, China's cultivated land area in the early 1980s was between 132.5–139.7 million hectares. Third, the satellite remote sensing data of the Ministry of Land and Resources of China after 1996 is basically credible. Corrections to the statistical data are based on the research results of Feng Zhiming *et al.* (2005). For details, please see Fig. 1.7.

In addition, Feng Zhiming *et al.* also pointed out that the restored data should only be used to reflect the trend of changes of China's cultivated land area during the corresponding periods and should not be used as the exact amount of cultivated land area at that time. Generally speaking, according to their corrected data, China's cultivated land area experienced fluctuations ever since 1949, with an overall increase before 1979, a slow decline after 1980s and a rapid decrease after 1999 due to the ecological conversion

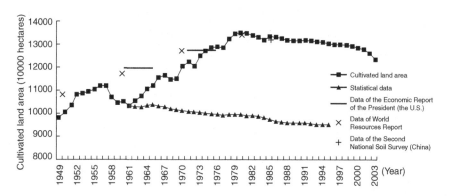

Figure 1.7. Corrections to the Changes of China's Cultivated Land Area.
Source: Feng Zhiming, Liu Baoqin and Yang Yanzhao (2005). A study of the changing trend of China's cultivated land area and data reconstructing: 1949–2003. *Journal of Natural Resources*, 20(1) 35–43.

of the cultivated land. We believe that the data correction made by Feng Zhiming *et al.* (2003) basically reflects the trend of changes of China's cultivated land area since the founding of New China. According to their research, the total area of cultivated land reached its peak in 1979, the very beginning of China's reform and opening-up, because the long-term policy of "taking grain as the key link" had promoted the continual increase of China's cultivated land area since the Cultural Revolution; whereas, the prosperous development of township enterprises after reform and opening-up had caused a continual decrease of China's cultivated land area since then. However, this trend of changes failed to adequately explain the possibility of an increase in the total area of cultivated land due to the fact that farmers' enthusiasm for agricultural production was greatly aroused by the implementation of the policy of the household contract responsibility system ever since China's reform and opening-up. In other words, the peak value of China's cultivated land area is more likely to occur after the 1980s.

To sum up, the changes of China's cultivated land area have been objectively discussed above. The data of *China Statistics Yearbook* was questioned by many scholars due to its defects in statistical methods, while the data of the Ministry of Land and Resources of China after 1996 has won the public's recognition due to improvements in measurement techniques and the advancement of statistical survey methods. Although *China Statistics Yearbook* provides the most complete data, there are many defects in this set of data, and the trend of changes shown by this set of data failed to explain the accurate data of surveying land and fixing production quota in 1953 and the widely-recognized data after 1996.

3.5. The trend of changes of China's cultivated land area per capita

As mentioned above, the data from *China Statistics Yearbook* before 1996 is erroneous and cannot be used for calculating the cultivated land area per capita. Therefore, China's cultivated land area per capita before 1996 is shown in a light-colored line in Fig. 1.8 for reference. After 1996, data of the Ministry of Land and Resources is used. In addition, the public has paid more attention to the changes in the total area of cultivated land in recent years because data after 1996 is more credible. China's cultivated land area

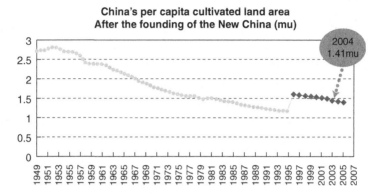

Figure 1.8.   Cultivated Land Area per Capita after the Founding of the New China.
Sources: Data of China's cultivated land area comes from *China Statistics Yearbook* (before 1996) and *China Agricultural Development Report* 2007 (after 1996), and data of population comes from Comprehensive Statistical Data Covering 50 Years Period of New China (1949– 1999) and *China Statistics Yearbooks*, summarized by the author. Please see the Appendix 3 for changes in population after the founding of the new China.

per capita was 1.41 mu in 2004 and 1.40 mu in 2005, showing that the downward trend has slowed down slightly.

4. Structural Changes of China's Cultivated Land Area

4.1. *Structural changes of China's cultivated land before the 1980s*

Reliability analyses of the historical data of this period mainly revolve around the structural changes of China's cultivated land area, and analyses were generally made from two aspects: The increase or decrease of cultivated land area.

Wasteland reclamation was the main means of cultivated land area increase. Bi Yuyun *et al.* (2000) sorted out and analyzed the area of wasteland reclamation during each five-year plan according to agricultural statistical materials. Please see Table 1.3 for details.

Before the 1980s, the main causes of the decrease of China's cultivated land area were as follows:

Table 1.3. Statistics of the Newly-added Cultivated Land Area in China Since the Founding of the New China (10,000 Hectares).

Period (Year)	Newly-added cultivated land area	Wasteland reclamation area
Recovery period (1950–1952)	—	—
First Five-Year Plan (1953–1957)	—	550
Second Five-Year Plan (1958–1962)	—	1,056
Adjustment period (1963–1965)	—	274
Third Five-Year Plan (1966–1970)	—	224
Fourth Five-Year Plan (1971–1975)	381	235
Fifty Five-Year Plan (1976–1980)	404	233
Sixth Five-Year Plan (1981–1985)	285	125
Seventh Five-Year Plan (1986–1990)	233	128
Eighth Five-Year Plan (1991–1995)	259	116

Source: Data compiled from relevant agricultural statistics.

Occupation of cultivated land for infrastructure construction. For instance, according to the estimation of National Bureau of Land Administration in *Research on the Overall Planning for National Land Use*, the total area of land for infrastructure construction in China was 10.684 million hectares in 1952 and 23.667 million hectares in 1980, an increase of 12.803 million hectares in 28 years.

Occupation of cultivated land for agricultural structural adjustment. Before the reform and opening-up, the agricultural structural adjustment in China went on rather slowly, and not much cultivated land was occupied. In addition, under the guidance of the policy of "taking grain as the key link", China's cultivated land was especially protected during that period. According to the statistics of the National Bureau of Statistics of China, the total area of garden land in China was 1.256 million hectares in 1952 and 4.543 million hectares in 1980, an increase of 3.287 million hectares in 28 years. In reality, conversion of cultivated land into garden land mainly occurred after the mid-1980s. Analysis of the data released by the National Bureau of Statistics between 1988 and 1995 shows that agricultural structural adjustment took up 62%, which was the largest proportion in the total loss of cultivated land area. Agricultural structural adjustment includes cultivated land converted into garden land, fishery ponds, forest land and grassland.

Besides, disaster destroyed cultivated land and deserted cultivated land due to soil erosion and desertification also led to the reduction of the cultivated land area.

4.2. *Structural changes of cultivated land from 1980s–1990s*

According to the statistics of the National Bureau of Statistics from 1988 to 1995, wasteland reclamation constituted 76%, the largest proportion, of the total increase in cultivated land during these eight years. Agricultural structural adjustment and reclamation accounted for 13% and 11% respectively. Among the various causes of loss of cultivated land, agricultural structural adjustment took up the biggest proportion of 62%, including cultivated land converted into garden land, fishery ponds, forest land and grassland; non-agricultural construction land took up the second largest share of 20%; natural disaster (water and wind erosion, sand deposit, floods) destroyed cultivated land accounted for 18%.

The newly-reclaimed cultivated land mainly came from regions with adverse natural conditions, including the northeastern region, the northwestern region and the southwestern region. Xinjiang, Yunnan, Heilongjiang, Inner Mongolia and Guangxi boasted the largest wasteland reclamation areas during the eight years, the total of which accounted for 60% of the total reclamation area in China. Interestingly, these provinces also had the largest disaster destroyed cultivated land areas. In terms of cultivated land occupation caused by agricultural structural adjustment, the eastern region and the central region had severe problems of conversion of cultivated land into garden land and fishery ponds, including Guangdong, Jiangsu, Liaoning, Hubei and Shanxi; provinces with the largest area of conversion of cultivated land into grassland included Inner Mongolia, Shaanxi, Xinjiang, Tibet and Yunnan, most of which were in the western region. Regions with serious problems of cultivated land occupation for non-agricultural construction were mostly distributed in the eastern coastal areas centered around the Pearl River Delta, the Yangtze River Delta, Beijing and Tianjin; other than these, there are similar serious problems in the central provinces, such as, Hubei, Henan, Anhui. It should be noted that the proportion of the non-agricultural construction land occupation of the 12 coastal provinces took up around

40% of the total area of non-agricultural construction land occupation from 1988 to 1991, and rose to 50%–55% in the four years that followed.

From 1988 to 1995, agricultural structural adjustment accounted for more than half of the total loss cultivated land area, including conversion of cultivated land back to grassland and conversion of cultivated land to garden land and fishery ponds. Conversion of cultivated land back to grassland was mainly driven by the government's policy of environmental and ecological protection and its ten major forestry ecological projects. Conversion of cultivated land to garden land and fishery ponds was mainly carried out by farmers driven by economic benefits. The area of garden land in China increased by four times from 1979 to 1996. Judging from the growth curve, this had a close relation with the increase of grain output. The two rapid expansion periods of garden land (1985–1988 and 1993–1996) occurred after the total grain output had risen to new levels. 1985–1988 was the fastest expansion period of garden land. This was because the per capita grain yield of farmers rapidly increased after the peak of grain output in 1983 and farmers began to look for new ways of land utilization for a higher income. Meanwhile, the per unit area yield of three major grain crops reached new heights around 1984, providing the possibility of agricultural diversification.

On an average, around 45,000 hectares of cultivated land were destroyed by disasters in China annually, which mainly happened in provinces of the northeastern region, the northwestern region and the southwestern region. Provinces with the largest disaster destroyed cultivated land areas were also those having the largest areas of wasteland reclamation. These are the regions with adverse conditions and fragile environments. Generally speaking, there are larger areas of disaster destroyed cultivated land in the northeastern region and the northwestern region, which are mainly threatened by desertification. In the southwestern region, mainly threatened by soil erosion, there are smaller areas of disaster destroyed cultivated land. However, there is significant loss in cultivated land productivity caused by soil erosion.

4.3. Structural changes of cultivated land since 1996

During this period, China's cultivated land area was supplemented through land consolidation, land reclamation, land exploitation and agricultural structural adjustment. According to the analysis of Tang Jian *et al.* (2006), the increase of cultivated land area through land consolidation, reclamation and exploitation stood at around 0.2 million to 0.3 million hectares, among which the largest increase of cultivated land area was achieved through land exploitation. For example, in 2003, land exploitation accounted for 69% of the total increase of cultivated land area through land consolidation, reclamation and exploitation. At the same time, the increase of cultivated land area through agricultural structural adjustment gradually dwindled down.

There are four major channels by which cultivated land is decreased, i.e., occupation for construction, disaster damage, ecological conversion, and agricultural structural adjustment. From Table 1.4 and Fig. 1.9, which summarized the relevant data of *China Agricultural Development Report 2007*, it is clear that ecological conversion of cultivated land is the main cause of cultivated land decrease in recent years, and in 2003 it alone accounted for almost 78% of the total decrease of cultivated land area. From 1998 to 2006, the proportions of the four channels by which cultivated land

Table 1.4. Structural Changes of Cultivated Land, 1998–2006 Unit: 1,000 Hectares.

Year	Cultivated land area decreased	Cultivated land occupation for construction	Disaster destroyed cultivated land	Ecological conversion of cultivated land	Agricultural structure adjustment
1998	570.4	176.2	159.5	164.6	70.1
1999	841.7	205.3	134.7	394.6	107.1
2000	1,566.0	163.3	61.7	762.8	578.2
2001	893.3	163.7	30.6	590.7	108.3
2002	2,027.4	196.5	56.4	1,425.5	349.0
2003	2,880.9	229.1	50.4	2,237.3	364.1
2004	1,146.0	145.1	63.3	732.9	204.7
2005	594.9	138.7	53.5	390.4	12.3
2006	582.8	167.3	35.9	339.4	40.2
Total	1,008,555.3	1,585.2	646	7,038.2	1,834
Percentage		14.3%	5.8%	63.4%	16.5%

Source: Compilation of data from *China Agricultural Development Report 2007*.

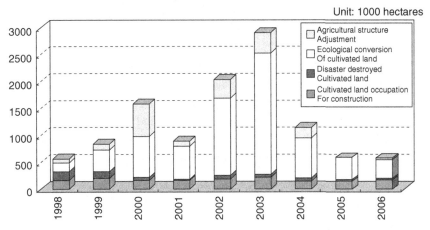

Figure 1.9. Structural Analysis of the Decrease of Cultivated Land, 1998–2006. *Source*: Compilation of data from *China Agricultural Development Report 2007*.

area is decreased are as follows: Occupation for construction 14.3%, disaster damage 5.8%, ecological conversion 63.4%, and agricultural structure adjustment 16.5%.

Zhao Qiguo *et al.* (2006) made a comparison of the decrease in cultivated land in the years between 1996 and 2004, and that between 1986 and 1995. From 1996 to 2004, decrease of cultivated land through agricultural structure adjustment was 1.81 million hectares, accounting for 17.16% of the total decrease, lower than that between 1986 and 1995; decrease due to occupation for construction was 1.652 million hectares, accounting for 15.68%, obviously higher than that between 1986 and 1995; decrease due to disaster was 6.032 million hectares, accounting for 5.73%, substantially lower than that between 1986 and 1995. Average annual loss was 75,400 hectares between 1996 and 2004, 42.82% of that between 1986 and 1995.

The rate of decrease of cultivated land area in recent years can be divided into three levels in terms of geographical distribution. The rate of decrease of Beijing, Shaanxi, Gansu and Inner Mongolia was the fastest at over 10%. These provinces were located in fragile ecological zones and the ecological conversion of cultivated land in these regions resulted in substantial reduction of cultivated land. The rate of decrease of the Yangtze River Basin and the Eastern coastal regions was in the second level, in the range of 2.2%–10%, the main causes of reduction are ecological conversion and occupation for construction. For regions with abundant potential resources of cultivated land, the rate of decrease was the lowest, i.e., below 1%. Among all the provinces in China, a net increase of cultivated land area is witnessed only in Xinjiang and Tibet. Provinces with the largest net decrease of cultivated land area included Inner Mongolia, Shaanxi, Sichuan, Shanxi, Hebei, Guizhou and Guangdong, all of which (with the exception of Guangdong) belong to the Western region and its fringe regions, characterized by fragile ecological environments, and their net decrease of cultivated land area accounted for 70% of the total. At the same time, the coastal regions and the middle and lower reaches of Yangtze River also witnessed a large total net decrease of cultivated land area, although these are regions with relatively good agricultural production conditions.

After a comprehensive analysis of the structural changes of China's cultivated land area, we arrived at a general conclusion: Before the reform and opening-up, the increase in cultivated land area mainly came from wasteland reclamation, and the main cause for the decrease of cultivated land area was occupation for construction; from the 1980s to the mid-1990s, the main cause for the decrease of cultivated land area was agricultural structure adjustment, and a prominent characteristic of this period was the dramatic increase of garden land area. From 1996 until now, the main means of increasing cultivated land area has been land exploitation, while the main cause of decrease in cultivated land area has been conversion of cultivated land to forests, and decrease of cultivated land area has been due to agricultural structure adjustment smaller than that from the 1980s to the mid-1990s, however, decrease due to construction was larger than that from the 1980s to the mid-1990s. In summary, while cultivated land occupation for construction remains a rather important factor for the decrease of cultivated land area, decrease of cultivated land area due to conversion to forests is higher than that due to occupation for construction.

5. The General Situation of China's Cultivated Land Quality

5.1. The current situation of China's cultivated land quality

According to the data of the Second National Soil Survey[3] (1979–1991), the high-yield cultivated land made up 21.55% of the total area of cultivated land, the medium-yield cultivated land accounted for 37.23%, and the low-yield cultivated land occupied 41.22%. The regional distribution was as follows: There was a higher proportion of the high-yield cultivated land in the North China region and the middle and lower reaches of Yangtze River; the Loess Plateau and the arid region of Northwest China took up the largest percentage of the medium-yield cultivated land, the South China region and the Southwest China region took up the second largest proportion, and the Qinghai Tibet plateau and the Northeast China region took up the third largest percentage. The first-class good quality cultivated land with no constraint factors accounted for 41.33% of the total area of cultivated land; the second-class medium quality cultivated land with certain constraint factors took up 34.55%; the third-class poor quality cultivated land with many constraint factors occupied 20.47%. (Please see Fig. 1.10 for details.) The potential resources of new cultivated land in China are generally of low quality and are located in remote areas, and it is very difficult to exploit and make use of these resources.

Since the founding of the New China, great progress has been made in the expansion of irrigation areas, control of water logging, treatment of saline and alkaline land and the control of soil erosion; it is undeniable that the quality cultivated land has been greatly improved. However, in recent years, cultivated land quality degradation in China has become fairly

---

[3]Two soil surveys had been carried out in China in 1958 and 1979 respectively. The third national soil survey started from 2005 and has not been completed. The Geographical Information System (GIS) is a new applied technology which has been used in the estimation of the total area and quality of cultivated land ever since the 1970s, including land resources investigation, land evaluation, land use planning, complex mapping, etc.

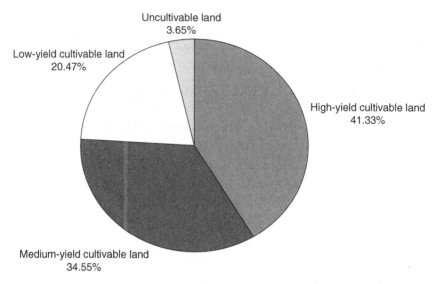

Statistical figure of cultivated land quality in China

Figure 1.10.   Statistical Figure of Cultivated Land Quality in China.
Source: 1: 1000000 Map of Chinese land Resource, p. 67. Beijing: China Renmin University Press.

severe, as manifested in soil erosion, soil desertification, soil productivity deterioration, and serious soil pollution caused by industrialization and urbanization (Zhao Qiguo *et al.*, 2006). Soil corrosion leads to land quality degradation, thinning of soil layer, deterioration of soil structure and quality, and loss of nutrients. It is estimated that about 5.0×10^9 tons of soil materials, 2.7×10^7 tons of soil organic matters, 5.5×10^6 tons of nitrogen, 6.0×10^3 tons of phosphorus, and 5.0×10^6 tons of potassium were lost annually due to soil erosion in China. However, no agreement has been reached as to the classification of soil deterioration and the grade division of deterioration. A rather convincing classification divided soil deterioration into four types, namely, soil deterioration caused by wind erosion, water erosion, physical factors, and chemical factors (Shen Weishou *et al.*, 2006). Based on the classification of previous authors, this report will carry out the assessment from the following three perspectives:

- *Soil sandification and desertification.* Apart from natural factors such as wind erosion and water erosion, there are also human factors. According to the *Bulletin of Status Quo of Desertification and Sandification in China* issued by State Forestry Administration in 2005, by the end of 2004, there was a total desertified land area of 2,636,200 square kilometers which made up 27.46% of the national territory, a total sandified land area of 1,739,700 square kilometers which accounted for 18.03% of the national territory, and the area of sandified cultivated land was 46,300 square kilometers which was equivalent to 2.66% of the total sandified land area. In terms of land use types which showed evident sandification trend, grassland took up the largest proportion of 68%, and cultivated land took up the second largest proportion of 23%. It is obvious that cultivated land was still severely threatened by sandification. Besides, soil secondary salinization caused by improper irrigation was rather serious in the Northwest China, North China and Northeast China, and aridization caused by the decrease of soil moisture due to over-exploitation of underground water was fairly common in Northwest China.

- *Soil deterioration of cultivated land.* According to the comprehensive evaluation of soil fertility, most of the soil in the red soil hilly region of Eastern China suffered different degrees of fertility deterioration, the fertility of the soil in this region belonged to medium and low fertility levels. The high-fertility soil, medium-fertility soil, and low-fertility soil accounted for 25.9%, 40.8% and 33.3% of the total cultivated land area in this region respectively. Soil fertility deterioration was very severe in the hilly areas of Guangdong province, Baise Area of Guangxi Zhuang Autonomous Region, Jiangxi Jitai Basin and the Southern area of Fujian Province. Besides, extensive tillage and soil management in a careless way, as well as soil erosion, have caused soil fertility deterioration in farmland and pasture, especially losses in soil organic matter and organic nitrogen. This problem is rather common all over the country and is most serious in Huang-Huai-Hai Plain, Songliao Plain and Inner Mongolia Grassland.

- *Land pollution caused by urbanization and industrialization.* Industrial and domestic wastewater discharge, the increase of coal burning, and extensive smelting of sulfide minerals resulted in environmental pollution and acid precipitation which in turn led to environmental acidification, including air acidification, soil acidification and water acidification, thus causing serious land pollution.

5.2. Impacts of cultivated land use changes on cultivated land quality

Land bio-productivity gives a good representation of the quality of cultivated land. Besides, in the Global Assessment of Soil Degradation (GLASOD), the degree of soil deterioration was classified according to the deterioration soil suitability for agricultural production, i.e., the decline of productivity and biological functions (Shen Weishou *et al.*, 2006).

A general practice to assess the biological productivity of a piece of land is to make an assumption on the type of the crop or crop combinations which could be grown on each land plot, and then to assess its biological productivity according to the following criteria: Acceptable output, land conversion, intensiveness of land utilization taking into consideration the constraints of technology, management and investment, etc.

Deng Xiangzheng *et al.* (2005) calculated the impact of cultivated land use changes on land bio-productivity by making use of the Agro-Ecological Zones (AEZ) methodology (developed by the Food and Agriculture Organization of the United Nations (FAO) in collaboration with the International Institute for Applied Systems Analysis (IIASA)) and the 1986–2000 Land Utilization Types (LUTs) dataset developed by the Chinese Academy of Sciences.

It was found that conversion of cultivated land to other uses resulted in an average decline of 2.2% in bio-productivity between 1986 and 2000, in which the bio-productivity of new cultivated land converted from other uses was about 5.08 million kcal/ha, while that of cultivated land converted to other uses was about 11.38 million kcal/ha. The impact of conversion of cultivated land on China's food security between 1986 and 2000 was minimal. China's cultivated land area increased by 1.9% over the study period, while the average bio-productivity dropped by merely 2.2%. The increase in cultivated land area could almost make up for the decline in bio-productivity caused by conversion of cultivated land. Hence, the author believed that the decline in land bio-productivity was negligible. The conclusion of this research that there was a decline in the quality of cultivated land was consistent with the trend shown in statistical data.

6. Conversion of Cultivated Land to Forests Program

Conversion of cultivated land to forests caused frequent and substantial changes in China's cultivated land area. What follows is an analysis of the background of this policy, its impact on land quality changes, regional distribution of this program and an assessment of its impact.

6.1. *The background of conversion of cultivated land to forests program*

Before the implementation of conversion of cultivated land to forests Program in 1999, China had adhered to the fundamental policy of "taking grain as the key link" for a long time. Although certain practices of converting cultivated land back to forests had been carried out since the founding of the New China, there was a lack of integration and continuity in the policy implementation. Over 70% of China's sloping cultivated land over 25° concentrated in the Western region. Due to the long-term destruction of forests and grassland and excessive land exploitation, the problem of land erosion in Central and Western region is the most severe in China. Ecological protection and restoration have become extremely urgent, especially after the disastrous flood of 1998.

Conversion of Cultivated Land to Forests Program is one of the six key forestry projects implemented in succession since the year of 1998. The six key forestry projects include: Natural Forest Protection Program, Conversion of Cultivated Land to Forests Program, Forest Shelterbelt Program (for Areas in the Vicinity of Beijing and Tianjin, and the Yangtze River Basin), Wildlife Conservation and Natural Reserves Development Program, and Fast-growing and High-yielding Timber Plantations Program. Among other things, conversion of Cultivated Land to Forests Program includes two aspects: Conversion of cultivated land to forests and grassland and the afforestation or planting of grass in suitable mountainous areas and plains.

In 1999, pilot work of conversion of cultivated land to forests was carried out in Sichuan, Shaanxi, and Gansu in line with the policy of

"Converting cultivated land to forests or pastures, prohibiting lumbering and promoting reforestation, giving relief to local residents in the form of grains, and allocating plots of land to individual contractors". In 2000, the State Council held meetings regarding Western Region Development and conversion of cultivated land to forests were listed as an important component as well as the foundation and breakthrough point of the Great West Exploitation Program. In September 2000, the State Council issued a directive *Several Opinions of the State Council on Further Improving the Pilot Work of Converting Cultivated land to Forest and Pasture.* In March 2001, conversion of cultivated land to forests was officially listed in China's 10th Five-Year Plan for National Economic and Social Development. After the implementation of pilot projects from 1999 to 2001, full implementation of Conversion of Cultivated Land to Forests Program started in the year 2002, covering an area of 25 provinces (including autonomous regions, municipalities) and involving the Xinjiang Production and Construction Corps. In this program, about 30 million rural households and 120 million farmers in total were involved. According to the statistics of the Office for Converting Farmland to Forest of the State Forestry Administration, by the year 2005, total area of cultivated land involved in the afforestation program was 8.997 million hectares and the afforestation of mountains and plains suitable for afforestation was 13.946 million hectares. By the end of 2006, the total area of cultivated land conversion was 364 million mu, among which afforestation of cultivated land was 139 million mu, afforestation area of deserted mountains and plains was 205 million mu, and 20 million mu of hills was closed off for afforestation. The forest coverage within the program implementation region rose by over 2%. The central government had invested over 130 billion RMB in total in this program and an average subsidy of 3500 RMB[4] had been provided to each household who had converted their cultivated land to forests.

---

[4]http://politics.people.com.cn/GB/1026/6363132.html www.people.com.cn [accessed on 11 October 2001].

6.2. Progress of conversion of cultivate land to forests program

6.2.1. Pilot demonstration stage, 1999–2001

In accordance with the natural characteristics of pilot sites and areas, the pilot projects were mainly carried out in five key regions, namely, (1) 54 counties from the mountainous regions of Hubei, Chongqing and Sichuan, (2) 45 counties from the Loess Hilly and Gully Region, 30 counties from the Yunnan–Guizhou Plateau region, (3) 16 counties from the Southwest Mountain and Ravine region, (4) 13 counties form the semi-arid regions of Inner Mongolia, Ningxia and Shaanxi, and (5) three counties in the cold pastures and steppes around the headwaters of the Yangtze River and the Yellow River (see Fig. 1.11).

Figure 1.11. Map of Distribution of the Pilot Demonstration Counties Involved in the Conversion of Cultivated Land to Forests and Grassland Program.
Source: Lu Dadao (2000). *China's Regional Development 2000.* p. 226.

During the three-year pilot demonstration period, the total area of cultivated land involved in afforestation was 1.162 million hectares, and the afforestation of desolate mountains and plains suitable for afforestation was 1.001 million hectares; 20 provinces, autonomous regions, municipalities, 400 counties, banners and county-level cities, 5,700 towns and townships, 27,000 villages, 4.1 million farmer households, and 16 million farmers were involved in total.

6.2.2. Program implementation stage, 2001–2010

Conversion of Cultivated Land to Forests Program was fully implemented in the year 2002. The implementation scope was expanded to 25 provinces (autonomous regions, municipalities) and 1897 counties (cities, county-level cities and banners) of the Xinjiang Production and Construction Corps. The program was implemented in two phases: 2001–2005 and 2006–2010. After ten years of implementation, the total area of cultivated land involved in conversion to forests was 14.667 million hectares, among which the sloping cultivated land with an inclination of over 25° was converted to forests and grassland, and the sandified cultivated land converted to forest was 2.667 million hectares or 38.9% of the total sandified cultivated land area; the afforestation area of desolate mountains and plains suitable for afforestation was 17,333 million hectares (see Table 1.5).

Since the pilot work of conversion of Cultivated Land to Forests Program was carried out in 1999, the State Council had issued four policies and regulations[5] on conversion of cultivated land to forests. Among others, the Regulations on Conversion of Farmland to Forests which was issued on 14 December 2002, marked the climax of conversion of Cultivated Land to

---

[5]The first one is *Several Opinions of the State Council on Further Improving the Pilot Work of Converting Cultivated land to Forest and Pasture* (Guo Fa, 2000, No. 24) which was issued on September 10, 2000. The second one is *Several Opinions of the State Council on Further Improving Policies and Measures of Converting Cultivated Land to Forests* (Guo Fa, 2000, No. 10) which was issued on April 11, 2002. The third one is the *Regulations on Conversion of Farmland to Forests* which was issued on December 14, 2002. The fourth one is the *Notice of the State Council on Improving Policies of Converting Cultivated Land to Forest* (Guo Fa, 2007, No. 25) which was issued on August 9, 2007.

Table 1.5. Task Planning of the Project of Converting Cultivated Land to Forests. (Unit: 10,000 Hectares).

Item		Total		Conversion of cultivated land to forests			Afforestation in the desolate mountains and plains suitable for afforestation		
	Subtotal	Yangtze river basin and southern regions	Yellow river basin and northern regions	Subtotal	Yangtze river basin and southern regions	Yellow river basin and northern regions	Subtotal	Yangtze river basin and southern regions	Yellow river basin and northern regions
Total Subtotal	3,200.0	1,505.3	1,694.7	1,466.7	754.2	712.5	1,733.3	751.1	982.2
Ecological forests	2,560.0	1,204.3	1,355.7	1,173.4	603.4	570.0	1,386.7	600.9	785.8
Cash forests	640.0	301.1	338.9	293.3	150.8	142.5	346.6	150.2	196.4
First five year Subtotal	1,533.3	693.9	839.4	666.7	322.1	344.5	866.7	371.8	494.9
Ecological forests	1,226.7	555.1	671.5	533.3	257.7	275.6	693.3	297.4	395.9
Cash forests	306.7	138.8	167.9	133.4	64.4	68.9	173.4	74.4	99.0
Second five year Subtotal	1,666.7	811.4	855.3	800.0	432.1	367.9	866.7	379.3	487.3
Ecological forests	1,333.3	649.1	684.2	640.0	345.7	294.3	693.3	303.5	389.9
Cash forests	333.3	162.3	171.1	160.0	86.4	73.6	173.4	75.8	97.4

Source: Li Shidong. *Research on the Conversion of Cultivated land to Forests in China*, p. 29.

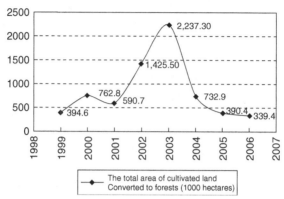

Figure 1.12. The Trend of the Total Area of Cultivated Land Converted to Forests. *Source: China Agricultural Development Report 2007,* compiled by the author.

Forests Program in China and signified that this program had embarked on a track toward institutionalization and standardization.

From Fig. 1.12, we can see rather clearly the trend of changes in the total area of cultivated land converted to forests in China. By the year 2003, the total area of cultivated land converted to forests reached its peak of 2.2373 million hectares. Then the state suddenly cut the target for conversion of cultivated land substantially from 60 million mu to 10 million mu in the year 2004 without any prior notice. The reason might be that the policy of converting cultivated land to forests was initiated in the late 1990s when an overall overproduction of food occurred in China, but large amounts of food stocks were consumed ever since the implementation of the policy, and there was a sharp rebound in the food price in the year 2003 due to the decreased food production and reduced food stocks. The Notice of the State Council on *Improving Policies of Converting Cultivated Land to Forests* (Guo Fa, 2007, No. 25) issued on August 9, 2007 adjusted the planning of converting cultivated land to forests. Apart from the four million mu which had already been allocated in the year 2006, the afforestation of the rest of the 20 million mu of cultivated land in the 11th Five-Year Plan was postponed. According to this notice, the State Council requested relevant departments to further find out the actual situation of cultivated land with a slope of over 25°, and to formulate plans for the implementation of converting cultivated land to forests in a practical and realistic way based on thorough investigations and careful experience summarization.

6.3. The impacts of conversion of cultivated land to forests and grassland on food security

An initial analysis made by Deng Xiangzeng *et al.* (2005) on the effects of conversion of cultivated land to forests and grassland showed that it had a minimal impact on food production and food price. For one thing, the quality of the cultivated land converted to forests and grassland was rather low with its average per mu yield being only 30% or even lower compared to the average level of the country. For another, farmers will input more on other cultivated land with better production conditions after they have converted some sloping cultivated land, and the increase in the output of cultivated land with better production conditions would make up for the decrease in the output caused by conversion of sloping cultivated land to a certain degree.

Dong Mei and Zhou Funing further analyzed the impacts of conversion of cultivated land to forests on food security. They believed that conversion of cultivated land to forests was not the main reason for the reduction in grain output. First, most of the cultivated land planned to be converted was sloping cultivated land and sandified land which had poor soil moisture and fertilizer conditions, and serious soil erosion; their grain output was low and unstable. It was estimated that the average per mu yield of China's sloping cultivated land over 15° was merely 236 jin (1 kg = 2 jin), and local farmers could hardly get enough seeds back from these sloping cultivated land when there was drought or other natural disasters. Even if the planned 220 million mu of sloping cultivated land and sandified land were converted to forests and grasslands, there would be a total decrease of 52 billion jin in grain output annually, constituting 5.2% of the average annual grain output during the Ninth Five-Year Plan (*The State Overall Planning for Conversion of Farmland to Forests, 2002*). Second, among all factors leading to the decrease in the grain output between 1998 and 2002, grain price and natural disasters ranked first, agricultural structural adjustment which caused the decrease in the sown areas and per unit area yield ranked second, and the implementation of the state policy of converting cultivated land to forests ranked third (Tan Jingrong, 2003). Third, it is estimated that by the year 2010, conversion of cultivated land to forests in the Western region would result in a reduction of four million tons in grain output while the increase in per unit yield would lead to an increase of ten million tons in grain output, resulting in a net increase of grain output of six million tons. Therefore, conversion of the marginal land would not threaten regional food security or food security of our country (Feng Zhiming, 2005).

Appendix

Appendix 1. Changes of China's Cultivated Land Area Since Its Establishment.

Year	Cultivated areas (1,000 hectares)	Year	Cultivated areas (1,000 hectares)
1949	97,881.33	1978	99,389.33
1950	100,356.00	1979	99,498.00
1951	103,671.33	1980	99,305.33
1952	107,918.67	1981	99,037.33
1953	108,528.67	1982	98,606.67
1954	109,354.67	1983	98,359.33
1955	110,156.67	1984	97,853.7
1956	111,824.67	1985	96,846.3
1957	111,830.00	1986	96,229.9
1958	106,900.67	1987	95,888.7
1959	104,579.33	1988	95,721.8
1960	104,861.33	1989	95,656
1961	103,310.67	1990	95,672.9
1962	102,903.33	1991	95,653.6
1963	102,726.67	1992	95,425.8
1964	103,312.00	1993	95,101.4
1965	103,594.00	1994	94,906.7
1966	102,958.00	1995	94,973.9
1967	102,564.00	1996	130,039.2
1968	101,553.33	1997	130,039.2
1969	101,460.00	1998	130,039.2
1970	101,134.67	1999	130,039.2
1971	100,699.33	2000	130,039.2
1972	100,614.67	2001	130,039.2
1973	100,212.67	2002	130,039.2
1974	99,912.00	2003	130,039.2
1975	99,708.00	2004	130,039.2
1976	99,388.67	2005	130,039.2
1977	99,246.67	2006	130,039.2
		2007	121,735.2

Source: *China Statistical Yearbook.*

Note: The cultivated areas data in *China Statistical Yearbook* showed the figure for October 31 1996 and remained the same after that. The source of this figure came from the *Communiqué on the General Results of Land Use Survey* which was issued jointly by the Ministry of Land and Resources, Nation Bureau of Statistics, and National Agricultural Census Office of China.

Appendix 2. Changes of China's Cultivated Land Area, 1978–1995. (Unit: 1,000 Hectares)

Year	Cultivated area (year-end)	Paddy fields	Dry fields	Decrease in cultivated area	Infrastructure construction	Village collective construction	Peasant housing construction
1978	99,389.5	25,419.7	73,969.8	800.9	144.5		
1980	99,305.2	25,322.2	73,983.0	940.8	97.7		
1985	96,846.3	25,033.0	71,813.3	1,597.9	134.3	92.3	97.0
1986	96,229.9	25,055.1	71,174.7	1,108.3	109.6	58.5	84.5
1987	95,888.7	25,104.1	70,784.5	817.5	104.6	52.0	57.5
1988	95,721.8	25,077.7	70,644.1	644.7	87.8	37.4	37.6
1989	95,656.0	25,265.8	70,390.2	517.5	70.1	34.6	27.4
1990	95,672.9	25,518.9	70,154.1	467.4	66.3	30.3	36.7
1991	95,653.6	25,706.5	69,947.1	488.0	71.9	33.4	20.5
1992	95,425.8	25,597.2	69,828.6	738.7	131.7	64.1	23.9
1993	95,101.4	25,028.0	70,073.4	732.3	161.0	86.0	24.0
1994	94,906.7	24,762.9	70,143.8	708.7	132.6	80.2	33.0
1995	94,970.9	24,850.5	70,120.4	621.1	111.9	84.9	31.6

Source: China Statistical Yearbook.

Note: It was indicated in *China Statistical Yearbook* that "figures for the cultivated areas are underestimated and must be further verified".

Appendix 3. China's Population Changes Since the Founding of the New China.

Year	Population (year-end) (10,000 persons)	Year	Population (year-end) (10,000 persons)	
1949	54,167	1978	96,259	
1950	55,196	1979	97,542	
1951	56,300	1980	98,705	
1952	57,482	1981	100,072	
1953	58,796	1982	101,654	
1954	60,266	1983	103,008	
1955	61,465	1984	104,357	
1956	62,828	1985	105,851	
1957	64,653	1986	107,507	
1958	65,994	1987	109,300	
1959	67,207	1988	111,026	
1960	66,207	1989	112,704	
1961	65,859	1990	114,333	
1962	67,295	1991	115,823	
1963	69,172	1992	117,171	
1964	70,499	1993	118,517	
1965	72,538	1994	119,850	
1966	74,542	1995	121,121	
1967	76,368	1996	122,389	
1968	78,534	1997	123,626	
1969	80,671	1998	124,810	124,761
1970	82,992	1999	125,909	125,786
1971	85,229	2000		126,743
1972	87,177	2001		127,627
1973	89,211	2002		128,453
1974	90,859	2003		129,227
1975	92,420	2004		129,988
1976	93,717	2005		130,756
1977	94,974			

Source: *Comprehensive Statistical Data Covering 50 years Period of New China (1949–1999)* and *China Statistical Yearbook*.

Note: The population data in the *Comprehensive Statistical Data Covering 50 years Period of New China* terminated in the year 1999; the population data for 1998 and 1999 in *China Statistical Yearbook* were different from that in the *Comprehensive Statistical Data Over 50 years Period of New China*.

Appendix 4. China's Cultivated Land Area (1983–2006). (Unit: 1,000 Hectares)

Year	Cultivated areas (year-end)	Increase in cultivated area	Decrease in cultivated area	Cultivated land occupation for construction	Disaster destroyed cultivated land	Biological conversion of cultivated land	Agricultural structural adjustment	Net decrease in cultivated area
1983	98,359.6		768.0	71.2	86.5			
1984	97,853.7	1,077.0	1,582.9	99.6	153.7			505.9
1985	**96,846.3**	**590.5**	**1,597.9**	**134.3**	**92.3**	**97.0**		**1,007.4**
1986	96,229.9	491.9	1,108.3					616.4
1987	95,888.7	476.3	817.5					341.2
1988	95,721.8	477.8	644.7					166.9
1989	95,656.0	451.7	517.5					65.8
1990	**95,672.9**	**484.3**	**467.4**					**−16.9**
1991	95,653.6	468.7	488.0					19.3
1992	95,425.8	510.9	738.7					227.8
1993	95,101.4	408.0	732.4					324.4
1994	94,906.7	513.9	708.6					194.7
1995	**94,973.9**	**686.7**	**621.0**					**−65.7**
1996	130,039.2							
1997	129,903.1							136.1

(Continued)

Appendix 4. (*Continued*)

Year	Cultivated areas (year-end)	Increase in cultivated area	Decrease in cultivated area	Cultivated land occupation for construction	Disaster destroyed cultivated land	Biological conversion of cultivated land	Agricultural structural adjustment	Net decrease in cultivated area
1998	129,642.1	309.4	570.4	176.2	159.5	164.6	70.1	261.0
1999	129,205.5	405.1	841.7	205.3	134.7	394.6	107.1	436.6
2000	**128,243.1**	**603.7**	**1,566.0**	**163.3**	**61.7**	**762.8**	**578.2**	**962.4**
2001	127,615.8	265.9	893.3	163.7	30.6	590.7	108.3	627.3
2002	125,929.6	341.2	2,027.4	196.5	56.4	1,425.5	349.0	1686.2
2003	123,392.2	343.5	2,880.9	229.1	50.4	2,237.3	364.1	2537.4
2004	122,444.3	345.6	1,146.0	145.1	63.3	732.9	204.7	800.3
2005	**122,082.7**	**306.7**	**594.9**	**138.7**	**53.5**	**390.4**	**12.3**	**361.6**
2006	121,800.0	367.2	582.8	167.3	35.9	339.4	40.2	266.7
	121,775.9							

Source: China Agricultural Development 2007.

Note: Cultivated area data from 1986 to 1995 comes from the *Statistical Communiqué of the People's Republic of China on National Economic and Social Development* by the National Bureau of Statistics; cultivated area data in and after 1996 was adapted in accordance with the *Communiqué on Land and Resources of China* by the Ministry of Land and Resources. Through land market rectification, it was found that the total area of cultivated land used for construction but their change reports were not submitted was 147,700 thousand hectares in 2004, 73,400 hectares in 2005 and 91,200 hectares in 2006.

Appendix 5. Changes of China's Cultivated Land Area (1961–2005). (Unit: 1,000 Hectares)

Year	National territorial area	Land area	Agricultural area	Arable area and lasting crops area	Arable area	Lasting crops	Lasting meadows and pastures	Forest area	Other area	Inland water
1961	959,805	932,745	343,248	105,248	103,397	1,851	238,000			27,060
1962	959,805	932,745	346,001	105,001	103,100	1,901	241,000			27,060
1963	959,805	932,745	348,854	104,854	102,903	1,951	244,000			27,060
1964	959,805	932,745	351,706	104,706	102,705	2,001	247,000			27,060
1965	959,805	932,745	355,509	104,509	102,443	2,066	251,000			27,060
1966	959,805	932,745	359,112	104,112	101,981	2,131	255,000			27,060
1967	959,805	932,745	362,715	103,715	101,494	2,221	259,000			27,060
1968	959,805	932,745	366,313	103,313	101,012	2,301	263,000			27,060
1969	959,805	932,745	370,928	102,928	100,547	2,381	268,000			27,060
1970	959,805	932,745	375,518	102,518	100,057	2,461	273,000			27,060
1971	959,805	932,745	380,165	102,165	99,619	2,546	278,000			27,060
1972	959,805	932,745	384,781	101,781	99,155	2,626	283,000			27,060
1973	959,805	932,745	390,388	101,388	98,682	2,706	289,000			27,060
1974	959,805	932,745	397,029	102,029	99,243	2,786	295,000			27,060
1975	959,805	932,744	401,637	100,637	97,766	2,871	301,000			27,061
1976	959,805	932,744	407,590	100,590	97,639	2,951	307,000			27,061
1977	959,805	932,743	413,532	100,532	97,501	3,031	313,000			27,062
1978	959,805	932,743	419,396	100,316	97,201	3,115	319,080			27,062
1979	959,805	932,743	426,423	100,423	97,215	3,208	326,000			27,062
1980	959,805	932,743	434,220	100,219	96,924	3,295	334,001			27,062

(Continued)

Appendix 5. *(Continued)*

Year	National territorial area	Land area	Agricultural area	Arable area and lasting crops area	Arable area	Lasting crops	Lasting meadows and pastures	Forest area	Other area	Inland water
1981	959,805	932,743	441,909	100,908	97,527	3,381	341,001			27,062
1982	959,805	932,742	449,200	101,199	97,723	3,476	348,001			27,063
1983	959,805	932,742	470,903	114,902	111,221	3,681	356,001			27,063
1984	959,805	932,742	478,901	115,900	111,914	3,986	363,001			27,063
1985	959,805	932,742	495,897	125,896	120,805	5,091	370,001			27,063
1986	959,805	932,742	504,997	126,996	120,800	6,196	378,001			27,063
1987	959,805	932,742	513,795	128,794	121,593	7,201	385,001			27,063
1988	959,805	932,742	522,704	129,703	122,189	7,514	393,001			27,063
1989	959,805	932,742	531,003	131,002	123,286	7,716	400,001			27,063
1990	959,805	932,742	531,398	131,397	123,678	7,719	400,001	157,141	244,203	27,063
1991	959,805	932,742	531,392	131,391	123,672	7,719	400,001	159,127	242,223	27,063
1992	959,805	932,742	531,784	131,783	123,762	8,021	400,001	161,113	239,845	27,063
1993	959,805	932,742	532,783	132,782	123,859	8,923	400,001	163,099	236,860	27,063
1994	959,805	932,742	533,480	133,479	123,956	9,523	400,001	165,085	234,177	27,063
1995	959,806	932,748	534,701	134,700	124,055	10,645	400,001	167,071	230,976	27,058

(Continued)

Appendix 5. (*Continued*)

Year	National territorial area	Land area	Agricultural area	Arable area and lasting crops area	Arable area	Lasting crops	Lasting meadows and pastures	Forest area	Other area	Inland water
1996	959,807	932,748	535,080	135,079	124,152	10,927	400,001	169,057	228,611	27,059
1997	959,807	932,748	535,372	135,371	124,143	11,228	400,001	171,043	226,333	27,059
1998	959,807	932,748	537,866	137,865	126,634	11,231	400,001	173,029	221,853	27,059
1999	959,807	932,748	544,962	144,961	133,630	11,331	400,001	175,015	212,771	27,059
2000	959,808	932,748	548,658	148,657	137,126	11,531	400,001	177,001	207,089	27,060
2001	959,808.6	932,748.6	555,156	155,155	143,624	11,531	400,001	181,058.8	196,533.8	27,060
2002	959,808.7	932,748.7	554,454	154,453	142,622	11,831	400,001	185,116.6	193,178.1	27,060
2003	959,808.7	932,748.7	553,401	153,400	141,119	12,281	400,001	189,174.4	190,173.3	27,060
2004	959,808.8	932,748.8	554,396	154,395	141,664	12,731	400,001	193,232.2	185,120.6	27,060
2005	959,808.8	932,748.8	556,328	156,327	143,296	13,031	400,001	197,290	179,130.8	27,060

Source: FAOSTAT.

Note: Cultivated land under lasting crops is land planted with perennial plants which need not be replanted after each harvest, such as cocoa, coffee and rubber. This category includes land grown with flowering shrubs, fruit trees, nut trees and vines, but excludes land under trees grown for wood or timber.

Chapter 2

The Unsuccessful Cultivated Land Protection System

1. Increasingly Enhanced Cultivated Land Protection System versus Increasingly Decreased Cultivated Land Area

1.1. Cultivated land protection system in China

To facilitate our logical description, we divide China's cultivated land protection policies into **direct policies, indirect policies and counteracting policies.** Direct policies refer to regulations which explicitly expound requirements on cultivated land protection and are made up of laws, regulations, policies and documents, such as *The Land Administration Law of the People's Republic of China, The Regulations on the Implementation of the Land Administration Law of the People's Republic of China,* and *The Regulations on the Protection of Basic Farmland*, etc. Indirect policies refer to regulations which play an indirect role in protecting cultivated land and are made up of *The Property Law of the People's Republic of China, The Law of the People's Republic of China on the Contracting of Rural Land, The Urban and Rural Planning Law of the People's Republic of China, Regulations Concerning Land Requisition for State Construction, Decision of the State Council on Furthering the Reform and Implementing Stricter Land Administration, Measures for the Administration of Preliminary Examination of the Land Used for Construction Projects, Measures on Disposal of Idle Land, Regulations on the Drafting and Examination of Overall Land Use Plans, Measures for the Examination of Overall Land Use Plans at Provincial Level, Measures for the Administration of Annual Plans on the Utilization of Land, Some Opinions on Land Development and Rehabilitation, Measures for Land Reserve Administration, Guiding Opinions on Improving Compensation for Land Acquisition and*

Resettlement System, Interim Measures for the Inspection and Acceptance of State-Invested Land Development and Consolidation Projects, Interim Measures for the Management of Implementation of State-Invested Land Development and Consolidation Projects, and the policies and measures implemented by the Ministry of Housing and Urban–Rural Development to crack down on housing of limited property rights, etc. Although these policies and measures were not issued for the sake of cultivated land protection, they may be regarded as indirect policies of cultivated land protection for the following reasons: they, theoretically, would enhance the collective ownership of land by the farmers, thereby would arouse the farmers' awareness of land protection and would protect their means in doing so; by attempting to put a rigid control on land use planning, theoretically, they would restrict the illegal occupation of cultivated land; by improving land acquisition procedures and increasing land acquisition costs, theoretically they would curb the blind urbanization process; by regulating and promoting land reclamation, theoretically they would increase the cultivated land area. Counteracting policies refer to policies which were issued for other political goals and results in the decrease of cultivated land area, such as, the policy of converting cultivated land to forests and grassland.

The above-mentioned laws and regulatory documents disclose the framework and means of cultivated land protection in China, which are manifested concretely as follows:

1.1.1. *Principles of direct policies*

I. The importance of cultivated land is explicitly made clear and special protection on cultivated land implemented.

For example, Article 4 (1) of *The Land Administration Law of the People's Republic of China (2004 Revision)* stipulates that "the state is to place a strict control on the usages of land". Article 4 (2) stipulates that "the state shall compile general plans to set usages of land including those of farm or construction use or unused. A strict control is to place on the turning of land for farm use to that for construction use to control the total amount of land for construction use and exercise a special protection on cultivated land".

II. Priority is given to conservation of total quantity, compensation of lost areas in other plots of land is allowed.

For example, Article 33 of *The Land Administration Law of the People's Republic of China (2004 Revision)* stipulates that "People's governments of all provinces, autonomous regions and municipalities shall implement strictly the general plans for the utilization of land and annual plans for the use of land, adopt measures to ensure not to reduce the total amount of cultivated land within their jurisdictions. Whereas reductions occur, the State Council shall order it to organize land reclamation within the prescribed time limit to make up for the reduced land in the same quantity and quality and the land administrative department of the State Council shall, together with agricultural administrative department, examine and accept it. Whereas individual provinces and municipalities find it difficult to reclaim enough land to make up for the land occupied due to scarce reserve resources, the total amount of land due to be reclaimed in their own regions may be reduced with the approval of the State Council but the rest of land for reclamation shall be made up for elsewhere".

III. The balance gap in land reclamation is allowed to be made up for by paying fees.

For example, Article 31(2) of *The Land Administration Law of the People's Republic of China (2004 Revision)* stipulates that "The State fosters the system of compensations to cultivated land to be occupied. In case of occupying cultivated land for non-agricultural construction, the units occupying the cultivated land should be responsible for reclaiming the same amount of land in the same quality as that occupied according to the principle of reclaiming the same amount of land occupied. Whereas units which occupy the cultivated land are not available with conditions of reclamation of land or the land reclaimed is not up to requirements, the units concerned should pay land reclamation fees prescribed by provinces, autonomous regions and municipalities for reclaiming land for cultivation".

IV. Administrative means — linking cultivated land protection with the performance evaluation of cadres

For instance, *The Regulations on the Protection of Basic Farmland* stipulates that "Local people's governments at all levels above the county level should include the work of protection of basic farmland in the national economic and social development plan as one of the contents of the goals of the government leadership responsibility system during the term of office and the implementation of which shall be supervised by the people's government at the next higher level. According to this regulation, local people's governments at all levels should establish the target and responsibility system for the protection of cultivated land, which focuses on basic farmland protection and the maintenance of dynamic balance in gross cultivated land area, and to conduct an annual assessment each year".

V. Economic means — land taxes and fees units

For example, Article 31 of *The Land Administration Law of the People's Republic of China (2004 Revision)* stipulates that "units which occupy the cultivated land are not available with conditions of reclamation of land or the land reclaimed is not up to requirements, the units concerned should pay land reclamation fees prescribed by provinces, autonomous regions and municipalities for reclaiming land for cultivation the land reclaimed"; Article 37 stipulates that "Construction work fails to start for over one year, land idling fees shall be paid according to the provisions by various provinces, autonomous region and municipalities"; Article 47 stipulates that "In expropriating vegetable fields in suburban areas, the units using the land should pay new vegetable field development and construction fund"; Article 55 stipulates that "Construction units that have obtained state-owned land by paid leasing can use the land only after paying the land use right leasing fees and other fees and expenses according to the standards and ways prescribed by the State Council". *The Interim Regulation of the People's Republic of China on Farmland Occupation Tax* stipulates that "any entity or individual who occupies any farmland to build a building or engage in non-agricultural construction shall be a taxpayer of farmland occupation tax. It (He) shall pay the farmland occupation tax under this Regulation". The tax and fee system regulated by laws is an important measure of protecting cultivated land through economic means.

VI. Legal means — deterrence through imposition of penalties

For example, Article 342 of *The Criminal Law of the People's Republic of China* stipulates that "whoever, in violation of the law or regulations on land administration, unlawfully occupies cultivated land and uses it for other purposes, if the area involved is relatively large and a large area of such land is damaged, shall be sentenced to fixed-term imprisonment of not more than five years or criminal detention and shall also, or shall only, be fined". Article 410 stipulates that "any functionary of a state organ who, engaging in malpractices for personal gain, violating the law and regulations on land administration and abusing his power, illegally approves the requisition or occupation of land or illegally transfers at low prices the right to the use of state-owned land, if the circumstances are serious, shall be sentenced to fixed-term imprisonment of not more than three years or criminal detention; if especially heavy losses are caused to the interests of the state or the collective, he shall be sentenced to fixed-term imprisonment of not less than three years but not more than seven years". Many laws and regulations, such as, *The Land Administration Law of the People's Republic of China, Regulations on the Implementation of the Land Administration Law of the People's Republic of China,* and *Regulations on the Protection of Basic Farmland* have stipulated the corresponding legal administrative responsibilities to be borne by anyone who violates relevant laws and regulations on cultivated land protection.

1.1.2. Means of realization of direct policies

I. The basic farmland protection system

Article 34 of *The Land Administration Law of the People's Republic of China (2004 Revision)* stipulates that "the state fosters the system of the basic farmland protection system". The basic farmland protection system includes the basic farmland protection responsibility system, the control system of land use in basic farmland protection areas, the strict examination and approval system of basic farmland occupation and the system of balancing occupation and reclamation, the basic farmland quality protection system, the basic farmland environment protection system, the supervision and inspection system of basic farmland protection, the examination and approval system of the conversion of agricultural land to other uses, etc. Article 44 stipulates that "whereas occupation of land for construction purposes involves the conversion of agricultural land into land for construction purposes, the examination and approval procedures in this regard shall be required. For projects of roads, pipelines and large infrastructure approved by the people's governments of provinces, autonomous regions and municipalities, land for construction has to be approved by the State Council where conversion of agricultural land is involved. Whereas agricultural land is converted into construction purposes as part of the efforts to implement the general plans for the utilization of land within the amount of land used for construction purposes as defined in the general plans for cities, villages and market towns, it shall be approved batch by batch according to the annual plan for the use of land by the organs that approved the original general plans for the utilization of land. The specific projects within the scope of land approved for conversion shall be approved by the people's governments of cities or counties. Land to be occupied for construction purposes other than those provided for in the second and third paragraphs of this article shall be approved by the people's governments of provinces, autonomous region and municipalities whereas conversion of agricultural land into construction land is involved".

II. The land development, consolidation and reclamation system

Article 38 of *The Land Administration Law of the People's Republic of China (2004 Revision)* stipulates that "the state encourages development of unused land by units or individuals according to the general plans for the utilization of land and under the precondition of protecting and improving the ecological environment, preventing water loss, soil erosion and desertification. Land suitable for agricultural use should have the priority of developing into land for agricultural use". Article 41 stipulates that "the state encourages land consolidation. People's governments of counties and townships (towns) shall organize rural collective economic organizations to carry out comprehensive consolidation of fields, water surface, roads, woods and villages according to the general plans for the utilization of land to raise the quality of cultivated land and increase areas for effective cultivation and improve the agricultural production conditions and ecological environment". Article 42 stipulates that "whereas land is damaged due to digging, cave-in and occupation, the units or individuals occupying the land should be responsible for reclamation according to the relevant provisions of the state; for lack of ability of reclamation or for failure to meet the required reclamation, land reclamation fees shall be paid, for use in land reclamation. Land reclaimed shall be first used for agricultural purposes".

1.1.3. *Indirect policies*

I. The right to contract and property rights

The Law of the People's Republic of China on the Contracting of Rural Land was adopted on August 2002. Article 10 of this Law stipulates that "the state protects the contract-undertaking party's right to transfer the operation of the contracted land lawfully, voluntarily, and for compensation"; Article 20 stipulates that "the duration of a contract for cultivated land shall be 30 years"; Articles 26 and 27 stipulate that within the duration of the contract, the party that lets the contract shall not withdraw or readjust the contracted land". The Property Law of the People's Republic of China classifies the right to contract as one of the property rights. These regulations grant farmers' collective ownership of land with stable expectations in terms of the duration, connotation, protection and usage of their rights to a certain degree, and provide farmers with conditions and legal space to enrich their means of exercising land rights as well as a basis for the protection of their land rights, thus playing a certain role in protection of cultivated land.

II. Urban and rural planning

The Urban and Rural Planning Law of the People's Republic of China was adopted at the 30th meeting of the Standing Committee of the Tenth National People's Congress of the People's Republic of China on October 28, 2007. By putting urban planning and village and town planning under unified administration, this law has certain defects in terms of legitimacy. Nevertheless, it exerts certain practical restrictive functions, thus playing the role of protecting cultivated land to certain extent.

(i) Giving priority to planning for the sake of cultivated land protection

In order to prevent certain local governments from expanding urban construction scale blindly, *The Urban and Rural Planning Law of the People's Republic of China* provides clear regulations on principles to be followed in the formulation and implementation of urban and rural planning. *The Urban and Rural Planning Law* stipulates that, in making and implementing urban and rural plans, attention shall be paid to following the principles of overall planning for urban and rural areas, rational layout, conservation of land, intensive development and planning before construction, to improving the ecological environment, promoting conservation and comprehensive utilization of resources and energy, to preserving cultivated land and other natural resources and historical and cultural heritage, to maintaining the local and ethnic features and traditional cityscape, to preventing pollution and other public hazards, and to meeting the need of regional population development, national defense construction, disaster prevention and alleviation, and public health and safety. Regarding specific circumstances of various kinds and at different levels, *The Urban and Rural Planning Law* stipulates that the people's governments at different levels or their competent departments of city planning administration should take the responsibility of urban system planning, overall urban planning, overall town planning, controlling detailed planning and construction detailed planning in accordance with their administrative power. Procedures of examination and approval of planning has also been clarified.

(ii) Planning according to law and accepting supervision

In order to enhance supervision over and inspection of the formulation, examination and approval, implementation and modification of urban and rural plans, *The Urban and Rural Planning Law of the People's Republic of China* devotes a full chapter for clearly stipulating the NPC supervision, administrative supervision and social supervision over the plans and constructions as well as various supervision measures.

The Urban and Rural Planning Law stipulates that the people's governments at or above the county level and the departments in charge of urban and rural planning under them shall improve supervision over and inspection of the formulation, examination and approval, implementation and modification of urban and rural plans.

It also stipulates that the local people's governments at various levels shall respectively report on the implementation of the urban and rural plans to the standing committees of the people's congresses at the same level or to the people's congresses of townships or towns, and shall subject themselves to supervision by the latter.

In order to facilitate supervision by the people, this law specially stipulates that the results of supervision and inspection and of the resolution of problems shall be published for public consultation and supervision.

(iii) Improve procedures, avoid casualness

The Urban and Rural Planning Law of the People's Republic of China stipulates that an urban or rural plan which is approved according to law shall provide the basis for administration of urban and rural development and planning, and it may not be modified without going through the statutory procedure.

It also stipulates that the authority in charge of the formulation of a provincial urban hierarchical plan or the overall plan of a city or town shall organize the relevant departments and experts to regularly assess the implementation of the plan, and solicit opinions from the public by holding appraisal conferences or hearings or by other means. The said authority shall submit the assessment report, attached with the opinions solicited, to the standing committee of the people's congress at the same level or the people's congress of the town and the original examination and approval authority.

(iv) Improvement on answerability and implementation of plans

In the light of the new features of the illegal activities in urban and rural plans and constructions, *The Urban and Rural Planning Law of the People's Republic of China* makes provisions to impose legal responsibilities on relevant parties, to impose more severe penalties for those who violate the law and discipline, to impose strict investigation on the legal responsibilities of relevant local people's governments and their administrative departments, and provides regulations on disposal of illegal buildings.

This law stipulates that, where an urban and rural plan should be formulated, as required by law, the authority concerned fails to take charge of such formulation, or fails to formulate, examine and grant approval, or modify an urban and rural plan in compliance with the statutory procedure, the people's government at a higher level shall order it to rectify and have it criticized in a circular; and it shall give sanctions, according to law, to the leading person of the people's government concerned and the other persons directly responsible.

It also stipulates that where the authority in charge of the formulation of urban and rural plans entrusts a unit which lacks the qualifications commensurate with the task, the people's government at a higher level shall order it to rectify and have it criticized in a circular; and the leaders of the people's government concerned and the other persons directly responsible shall be given sanctions according to law.

In order to effectively curb the construction of illegal buildings, *The Urban and Rural Planning Law* stipulates that where a unit engages in construction without obtaining the permit for a planned construction project or without complying with the provisions in the said permit, it shall be ordered to discontinue construction by the department in charge of urban and rural planning under the local people's government at or above the county level; if measures for rectification can be adopted to eliminate the impact on the implementation of the plan, it shall be ordered to make rectification within a time limit and be fined not less than 5% but not more than 10% the cost of the construction project; otherwise, it shall be ordered to demolish the project within a time limit; if the project cannot be demolished, the project itself or the unit's unlawful income shall be confiscated, and it may, in addition, be fined not more than 10% the cost of the construction project.

III. Land transfer

Standardization and rationalization of land transfer will determine the costs and profits sharing of cultivated land conversion. In this sense, the increasingly market oriented and legalized land transfer procedures would theoretically play the role of protecting cultivated land.

IV. Land requisition

Legalization of land requisition procedures and rationalization of compensation for land requisition will both protect the interests of farmers and determine the costs of cultivated land occupation.

V. Administrative licensing

To China, which is a country accustomed to realizing institutional goals through administrative means, the introduction of approval procedures and imposition of limitation on authority in making approval, are dependable means to protect cultivated land. This development also reflects the interests sharing and gaming among the central government and local governments at different levels.

(i) Basic regulations on the examination and approval of land uses in rural areas

The Land Administration Law of the People's Republic of China (2004 Revision) stipulates that "Land collectively owned by peasant shall be contracted out to members of the collective economic organizations for use in crop farming, forestry, animal husbandry and fisheries production under a term of 30 years. The contractees should sign a contract with the correspondent contractor to define each other's rights and obligations"; "Within the validity term of a contract, the adjustment of land contracted by individual contractors should get the consent from over two-thirds majority vote of the villagers' congress or over two-thirds of villagers' representatives and then be submitted to land administrative departments of the township (town) people's government and county level people's government for approval".

Land collectively owned by peasants may be contracted out to units or individuals who are not belonging to the corresponding collectives for farming, forestry, animal husbandry and fisheries operations; whereas a land collectively owned by peasant is contracted out for operation to ones not belonging to the corresponding collective organizations, a consent should be obtained from over two-thirds majority vote of the villagers' congress or over two-thirds of the villagers' representatives with the resulted contract being submitted to the township (town) people's government for approval. Land owned by the state may be contracted out to units or individuals for farming, forestry, animal husbandry and fisheries operations. The contractees and contractors should sign land use contracts to define each other's rights and obligations.

The land use right of peasant collectives shall not be leased, transferred or rented for non-agricultural construction, except in the case that an enterprise which has obtained a piece of land legally and conforms to the general plan for the utilization of land, and its right to use the land is being transferred according to the law due to bankruptcy or acquisition. In using the land for construction purposes defined in the general plan for the utilization of land of townships (towns) to start up enterprises or joint ventures together with other units or individuals by way of using land use right as shares, the rural collective economic organization shall file an application with land administrative departments of the local people's governments at and above the county level on the strength of documents of approval. The applications shall be approved by the local people's governments at and above the country according to the terms of reference provided for by various provinces, autonomous regions and municipalities whereas the use of land involving the occupation of agricultural land, the examination and approval procedures provided for in this law shall be followed. Land for construction purposes in starting enterprises shall be put under strict control. Provinces, autonomous regions and municipalities shall determine the standards for land use according to different trades and scale of operation of township enterprises.

One rural household can own one piece of land for building house, with the area not exceeding the standards provided for by provinces, autonomous regions and municipalities. The application for housing land after selling or leasing houses shall not be approved. Construction of rural houses should conform to the general plans for the utilization of land of townships (towns) and the original land occupied by houses and open spaces of villages should be used as much as possible for building houses. The use of land for building houses should be examined by the township (town) people's governments and approved by the county people's governments. Whereas occupation of agricultural land is involved the examination and approval procedure provided for in this law is required.

In using land for building public facilities and public welfare facilities, townships (towns) shall file an application with land administrative departments of local people's governments at and above the county level after being examined by the township (town) people's governments and the application shall be approved by the local people's governments at and above the county level according to the term of reference provided for by provinces, autonomous regions and municipalities. Where occupation of agricultural land is involved, the examination and approval procedures provided for this law are required.

(ii) Basic regulations on the examination and approval of land use planning

General plans for land use shall be examined and approved level by level. The national overall planning for land utilization shall be compiled by the competent department of land administration under the State Council in conjunction with the departments concerned under the State Council and submitted to the State Council for approval. Overall planning for land utilization of the provinces, autonomous regions and municipalities directly under the Central Government shall be compiled by the competent departments of land administration and other departments concerned at the same level under the organization of people's governments of the provinces, autonomous regions and municipalities directly under the Central Government and submitted to the State Council for approval. Overall planning for land utilization of municipalities that are seats of people's governments of the provinces and autonomous regions, municipalities with a population of over one million and municipalities designated by the State Council shall be compiled by the competent departments of land administration and other departments concerned at the same level under the organization of people's governments of the respective municipalities and submitted to the State Council for approval upon examination and consent of people's governments of the provinces and autonomous regions. Overall planning for other land utilization shall be compiled by the competent departments of land administration and other departments concerned at the same level under the organization of the people's governments concerned and submitted level by level to people's governments of the provinces, autonomous regions and municipalities directly under the Central Government for approval; among which, village (township) overall planning for land utilization shall be compiled by village (township) people's governments and submitted level by level to people's governments of the provinces, autonomous regions and municipalities directly under the Central Government or people's governments of municipalities with subordinate districts and autonomous prefectures authorized by people's governments of the provinces, autonomous regions and municipalities directly under the Central Government for approval.

Overall planning for land utilization should classify land into agricultural land, land for construction and unutilized land. County-level and village(township) overall planning for land utilization should, in accordance with requirements, delimit basic farmland protection zone, land reclamation zone, land for construction zone and reclamation prohibition zone, etc; among which, village(township) overall planning for land utilization should also, in the light of land use conditions, determine the use of each plot of land. Specific measures for land classification and delimitation of land utilization zones shall be worked out by the competent department of land administration under the State Council in conjunction with the departments concerned under the State Council. The planning duration of overall planning for land utilization shall generally be 15 years.

Revision of the general plans for land use shall be approved by the original organ of approval. Without approval, the usages of land defined in the general plans for the utilization of land shall not be changed. Whereas the purpose of land use defined in the general plans for the utilization of land needs to be changed due to the construction of large energy, communications, water conservancy and other infrastructure projects approved by the State Council, it shall be changed according to the document of approval issued by the State Council. If the purpose of land defined in the general plans for the utilization of land needs to be changed due to the construction of large energy, communications, water conservancy and other infrastructure projects approved by provinces, autonomous regions and municipalities, it shall be changed according to the document of approval issued by the provincial level people's governments if it falls into their terms of reference. Revision of overall planning shall be made by the original compiling organ in accordance with the approval document of the State Council or people's governments of the provinces, autonomous regions and municipalities directly under the Central Government. The revised overall planning for land utilization should be submitted to the original approval organ for approval. When the revised overall planning for land utilization at the next higher level involves revision of overall planning for land utilization at the next lower level, people's government at the next higher level shall notify people's government at the next lower level to make corresponding revision and submit it to the original approval organ for the record.

The annual plan for the land use shall be compiled in line with the national economic and social development program, the state industrial policies, general plans for land and the actual situation about the land for construction uses and the land utilization. The examination and approval procedures for the compilation of annual land use plans shall be the same as that for the general plans for land use. Once approved, they shall be implemented strictly.

(iii) Regulations on the examination and approval of land use for construction

Need of occupation of state-owned land for construction for a specific construction project within the scope of land for urban construction determined in the overall planning for land utilization shall be handled pursuant to the following provisions: (1) At the time of the construction project feasibility study, the competent department of land administration shall examine the matters relating to land use for the construction project and come up with a report on the preliminary examination of land use for the construction project; at the time of submission of the feasibility study for approval, the report on the preliminary examination of land use for the construction project produced by the competent department of land administration must be enclosed therewith. (2) The construction unit shall, on the strength of the relevant approval document of the construction project, file an application for land for construction with the competent department of land administration of municipal or county people's government, the competent department of land administration of the municipal or county government shall examine the same, draft a land provision plan and submit it to the municipal or county people's government for approval; where approval by people's government at the next higher level is required, it should be submitted to the people's government at the next higher level for approval. (3) Municipal or county people's government shall, upon approval of the land provision plan, issue a certificate of approval for land for construction for the construction unit. In the case of leasing of state-owned land, the competent department of land administration of municipal or county people's government shall conclude a contract on the leasing of state-owned land with the land user; in the case of appropriation for use of state-owned land, the competent department of land administration shall verify and issue a certificate of decision on the appropriation of state-owned land. (4) The land user should file an application for land registration according to law. For provision of land use right of state-owned land for construction in the form of tender or auction, the competent departments of municipal or county people's governments shall, in conjunction with the departments concerned, draw up a plan and submit it to the municipal or county people's government, the competent department of land administration of municipal or county people's government shall organize its implementation upon approval, and conclude a contract on the leasing of the land with the land user. The land user should file an application for land registration according to law.

Where a need arises for the occupation of state-owned unutilized land determined in the overall planning for land utilization for a specific construction project, it shall be handled pursuant to the provisions of the provinces, autonomous regions and municipalities directly under the Central Government; however, land use for key state construction projects, military installations and construction projects transcending the administrative areas of the provinces, autonomous regions and municipalities directly under the Central Government should be submitted to the State Council for approval.

Occupation of land for the implementation of urban planning within the scope of land for urban construction determined in the overall planning for land utilization shall be handled pursuant to the following provisions: (1) Municipal, county people's governments shall, pursuant to the annual plans for land utilization, draft agricultural land conversion plans, cultivated land supplement plans, land requisition plans, and submit them in batches and level by level to people's governments with the authority of approval. (2) The competent departments of land administration of people's governments with the authority of approval shall examine the agricultural land conversion plans, cultivated land supplement plans, land requisition plans, put forth examination remarks and submit the same to people's governments with the authority of approval for approval; among which, the cultivated land supplement plans shall be approved simultaneously with the approval of agricultural land conversion plans by people's governments that approve the agricultural land conversion plans. (3) Municipal, county people's governments shall, upon approval of the agricultural land conversion plans, cultivated land supplement plans and land requisition plans, organize their implementation and provide land separately according to specific construction projects. For occupation of land for the implementation of village and township planning within the scope of land for village and township construction determined in the overall planning for land utilization, municipal, county people's governments shall draft agricultural land conversion plans and cultivated land supplement plans and process them pursuant to the procedures prescribed in the preceding paragraph.

(iv) Regulations on conversion of agricultural land for other uses

To use land for a specific construction project, an application must be filed according to law for the use of state-owned land for construction within the scope of urban land for construction determined in the overall planning for land utilization. Where an actual need arises for the use of land outside the scope of land for urban construction determined by the overall planning for land utilization for such construction projects as energy, communications, water conservancy, mines and military installations involving agricultural land, it shall be handled pursuant to the following provisions: (1) At the time of the construction project feasibility study authentication, the competent department of land administration shall examine the matters relating to land use for the construction project and come up with a report on the preliminary examination of land use for the construction project; at the time of submission of the feasibility study for approval, the report on the preliminary examination of land use for the construction project produced by the competent department of land administration must be enclosed therewith. (2) The construction unit shall, on the strength of the relevant approval decumbent of the construction project, file an application for land for construction with the competent department of municipal or county people's government, the competent department of municipal or county people's government shall examine the application, draw up an agricultural land conversion plan, land requisition plan and land provision plan (where state owned agricultural land is involved, no land requisition plan shall be drafted), which shall, upon examination, verification and consent of the municipal or county people's government, be submitted level by level to the people's government with the authority of approval for approval; among which, the cultivated land supplement plan shall be simultaneously approved by the people's government that approves the agricultural land conversion plan at the time of approval of the agricultural land conversion plan; the land provision plan shall be simultaneously approved by the people's government that approves land requisition at the time of approval of the land requisition plan (where state-owned agricultural land is involved, the land provision plan shall be simultaneously approved by the people's government that approves the agricultural land conversion at the time of approval of the agricultural land conversion plan). (3) Municipal, county people's governments shall, upon approval of the agricultural land conversion plan, cultivated land supplement plan, land requisition plan and land provision plan, organize their implementation and issue the certificate of approval of land for construction to the construction unit. Where there is paid-for use of state-owned land, the competent department of land administration of municipal or county people's government shall conclude a contract on the paid-for use of state owned land with the land user; where state-owned land is appropriated for use, the competent department of land administration of municipal or county people's government shall verify and issue a certificate of decision on the appropriation of state-owned land to the land user. (4) The land user should file an application for land registration according to law. Where an actual need arises for land use for a construction project outside the scope of land for urban construction determined in the overall planning for land utilization involving unutilized land under peasants' collective ownership, only land requisition plan and land provision plan shall be submitted for approval.

For projects of roads, pipelines and large infrastructure approved by the people's governments of provinces, autonomous regions and municipalities, land for construction has to be approved by the State Council whereas conversion of agricultural land is involved. Whereas agricultural land is converted into construction purposes as part of the efforts to implement the general plans for the utilization of land within the amount of land used for construction purposes as defined in the general plans for cities, villages and market towns, it shall be approved batch by batch according to the annual plan for the use of land by the organs that approved the original general plans for the utilization of land. The specific projects within the scope of land approved for conversion shall be approved by the people's governments of cities or counties. Land to be occupied for other construction purposes shall be approved by the people's governments of provinces, autonomous region and municipalities whereas conversion of agricultural land into construction land is involved.

(v) Regulations on the examination and approval of land requisition

The requisition of the following land shall be approved by the State Council: (1) Basic farmland. (2) Land exceeding 35 hectares outside the basic farmland. (3) Other land exceeding 70 hectares. Requisition of land other than prescribed in the preceding paragraph shall be approved by the people's governments of provinces, autonomous regions and municipalities and submitted to the State Council for the record. Requisition of agricultural land should first of all go through the examination and approval procedure for converting agricultural land into land for construction purposes according to the provisions of this law. Whereas conversion of land is approved by the State Council, the land requisition examination and approval procedures should be completed concurrently with the procedures for converting agricultural land to construction uses and no separate procedures are required. Whereas the conversion of land is approved by people's governments of provinces, autonomous regions and municipalities within their terms of reference, land requisition examination and approval procedures should be completed at the same time and no separate procedures are required. Whereas the term of reference has been exceeded, separate land requisition examination and approval procedures should be completed according to the provisions.

(vi) Regulations on emergency land use and temporary land use

For emergency use of land required for dealing with an emergency or disaster relief, the land may be used first. Among which, where the land is for temporary use, the original state should be restored when the disaster is over and the land shall be returned to its original user for use, and formalities of examination and approval for land use are no longer required; when it falls into land for permanent construction, the construction unit should, within six months after the disaster is over, file an application for making up the formalities of examination and approval of land for construction.

In the case of temporary using of state-owned land or land owned by peasant collectives by construction projects or geological survey teams, approval should be obtained from the land administrative departments of local people's governments at and above the county level. Whereas the land to be temporarily used is within the urban planned areas, the consent of the urban planning departments should be obtained before being submitted for approval. Land users should sign contracts for temporary use of land with related land administrative departments or rural collective organizations or villagers committees depending on the ownership of the land and pay land compensation fees for the temporary use of the land according to the standard specified in the contracts. Users who use the land temporarily should use the land according to the purposes agreed upon in the contract for the temporary use of land and should not build permanent structures. The term for the temporary use of land shall not usually exceed two years.

(vii) Regulations on development and examination and approval of land whose use right has not been established

Whoever engages in development of state-owned barren hills, barren land or barren shoals the land use right of which has been established for cultivation, forestry, animal husbandry and fishery production in land reclamation zones determined by the overall planning for land utilization should file an application with the competent department of land administration of people's government above the county level of the locality wherein the land is located and submit the same to people's government with the authority of approval for approval. Whoever engages in single-time development of state-owned barren hills, barren land or barren shoals under 600 hectares the land use right of which has not been established shall be subject to the approval of local people's government above the county level pursuant to the limits of authority prescribed by the provinces, autonomous regions and municipalities directly under the Central Government; the case of development of over 600 hectares shall be submitted to the State Council for approval. Development of state-owned barren hills, barren land or barren shoals the land use right of which has not been established for cultivation, forestry, animal husbandry or fishery production may, upon approval of people's government above the county level, be assigned to development units or individuals for long-term use, and the longest duration of use shall not exceed 50 years. Units or individuals are prohibited from engaging in land development activities in reclamation prohibition zones determined by the overall planning for land utilization.

The most recent land examination and approval reform is actually heading towards centralization rather than the seeming decentralization. For instance, the State Council issued *The Circular of the State Council on the Problems Concerning Strengthening Land Regulation* in 2006 which stipulates that the method of examination and approval of the change of use of agricultural land as well as land requisition by the State Council on a batch-by-batch basis shall be changed into one whereby the people's government at the provincial level summarizes the situation and reports it to the Ministry of Land and Resources as well as the State Council for examination and approval on an annual basis. It was claimed that this move expanded the power of the people's government at the provincial level in the adjustment and arrangement of land use within the provinces and exercised strict control and management on the increase of urban construction land. However, this move also transfers the items which cities could originally examine by themselves to the provincial level. The applications for conversion of agricultural land to other uses and the implementation plans of land requisition of the cities, are firstly examined by the provincial departments of land and resources, which will be responsible for checking projects associated with allocated land, the scope of compensated land transfer, land survey and demarcation, adherence to the application procedure for approval before land requisition, the allotment of land requisition compensations, etc. They will also be responsible for the supervision and inspection of the implementation of the projects after they have been approved by the people's governments at the provincial level. More importantly, apart from submitting annual reports to the State Council, the people's governments at the municipal level will have to report to the people's governments at the provincial level when specific land use was involved. With the strengthened accountability system of land examination and approval imposed on the people's governments at the provincial level by the central government, this reform practically withdrew the power of land examination and approval.

At the same time, the land examination and approval system reform in 2006 clearly proposed to focus on strict control of the increase of construction land. It was claimed that the reform scheme will exert "all-round control" on the increase of construction land through controlling the following six aspects: Examination and approval links, examination and approval items, limitation in authority, methods of application for approval, procedures of application for approval and examination contents. Generally speaking, this reform focused on the following three aspects. First, the regulation on examination and approval was strengthened. "Unlike the examination and approval at different levels in the past, this reform requires strict procedures, complete coverage and whole-course management, and is mainly manifested by examination and approval of schemes and plans and the increase of construction land, which can be further divided into examination and approval of change of urban land uses, examination and approval of the transfer of rural house sites and agricultural land, etc". The implementation of regulation on examination and approval institutionally controlled the conversion of agricultural land for construction uses. Second, strict control was exerted on the examination and application for approval of the increase of construction land while ensuring the essential land for key national projects. Moreover, land examination and approval authority was further centralized by the central government and uniformly regulated by the state. A system coupling the examination and approval of construction land with the national industrial policies was established. Land use for projects of the industries restricted by the state, no matter how much it is, should be examined and approved by the State Council or examined and approved by the state land and resources departments authorized by the State Council; land use for projects of industries prohibited by the state should all be disapproved; the authority of examination and approval of the conversion of agricultural land for construction uses is revoked in regions where the per capita cultivated land area is below a certain limit. Third, exerting stricter examination and approval, i.e., stricter examination and approval was carried out on the overall land use planning, the balance of land occupation and reclamation, industrial policies, the conversion of agricultural land for other uses and land requisition in different regions. In addition, the Ministry of Land and Resources also exerted strict control on the total area of construction land through the land supply registration system, making public land supply information, supervision and inspection of the implementation of the overall land use planning, prohibition of further land supply after illegal land supply cases and clarification of the legal liabilities of various law-violating behaviors.

1.1.4. The counteracting policy of returning cultivated land to forests

China's grain production has been standing at a high position and its grain reserves has been expanding since 1995, with the sum of the state grain reserves and farmers' grain storage being equivalent to China's annual grain output (about 500 million tons). Against this background, the state issued the policy of returning cultivated land to forests for the following three purposes: Reducing the grain crop planted area and intervening in the grain market; protecting the environment; creating favorable conditions for the development and investment promotion of the Western regions. Besides, as China had 50 million mu of barren hills and slopes over 25° and the per mu grain yield of crops on barren hills and slopes was about 200 jin, it was calculated that the decrease of grain output due to conversion of cultivated land to forests and grassland would be about five million tons by assuming that the 50 million mu of barren hills and slopes could all be brought into production. According to the national grain subsidy standard for conversion of cultivated land to forests and grassland, 150 kg of grain are subsidized annually for each mu of converted cultivated land in the upper reaches of the Yangtze River, and 100 kg of grain are subsidized annually for each mu of converted cultivated land in the middle and upper reaches of the Yellow River. Therefore, five million tons of grain would be subsidized for the 50 million mu of barren hills and slopes in the grain subsidy scheme. In comparison with China's safe grain reserve of 20 million tons or 5% of the nation's average annual grain output and the import quota of around 20 million tons after China's entry into the WTO, the sum of five million tons of decrease in grain output due to conversion of cultivated land to forests and grassland and the grain subsidy of five million tons to farmers who converted cultivated land to forests and grassland obviously posed no threat to China's food supply.

However, China implements quantitative protection of cultivated land, and conversion of cultivated land to forests resulted in the reduction in cultivated land area anyhow.

1.2. What is wrong with the policies?

Although the strictest cultivated land protection policy is being implemented in China, including examination and approval, strict investigation, vertical management and land supervision and inspection, the cultivated land reserve planning has been repeatedly broken. In the second round of national overall planning for land use which used 1996 as the base year, the cultivated land reserve in 2000 was set at 1.94 billion mu and that in 2010 was set at 1.92 billion mu. However, the nation's cultivated land reserve had fell to 1.92356 billion mu at the very beginning to 2000, 16.35 million mu lower than the bottom line of 1.94 billion mu. In such cases, the target of 1.92 billion mu of cultivated land reserve in 2010 was forced to be advanced to 2005 at the start of the Tenth Five-Year Plan. However, only 1.831 billion mu of cultivated land was left by 2005 and 1.827 billion mu was left by 2006, approaching very close to the red line of 1.8 billion mu.

1.2.1. *Defaults in principles of cultivated land protection*

Under the principle of allowing land reclamation to be substituted by payment of fees, local governments are far more enthusiastic about collecting cultivated land reclamation fees than requesting the units occupying the cultivated land to reclaim wasteland. Article 31(2) of *The Land Administration Law of the People's Republic of China* stipulates that "the state fosters the system of compensations to cultivated land to be occupied. In the cases of occupying cultivated land for non-agricultural construction, the units occupying the cultivated land should be responsible for reclaiming the same amount of land in the same quality as that occupied according to the principle of reclaiming the same amount of land occupied. Whereas units which occupy the cultivated land are not available with conditions of reclamation of land or the land reclaimed is not up to requirements, the units concerned should pay land reclamation fees prescribed by provinces, autonomous regions and municipalities for reclaiming land for cultivation of the land reclaimed".

The administrative accountability system of cultivated land protection states that the higher level government should investigate the lower level government for its legal liabilities. However, legal liabilities put the emphasis on legal rights of the state and lack emphasis and regulations on legal rights of cultivated land right owners, as such is not conducive to protection of cultivated land.

1.2.2. *Local governments' gaming ability counteracts the effects of laws and policies of the central authorities (quoted from The Implementation and Protection of Property Rights of Land in the Process of Urbanization by Zhao Nong)*

Great differences and even conflicts exist between the central government and local governments in terms of the aims of land policies. Therefore, the game between the two parties was carried out under conflicting objectives. The central government pays more attention to protection of cultivated land and the farmers' situation because these two aspects will directly affect social stability and the image of the central government. Local governments, however, are more concerned with economic growth rates and revenues, both of which have something to do with land use and development. Certainly, the central government is also concerned with economic growth rates, but the degree of concern varies as economic development status changes. In other words, it pays more attention to economic growth in times of economic recession and less attention in times of economic overheating. Thus, the central government plays a role in marginal decrease of economic growth rates.

When the policy aims of the central government do not contradict with those of local governments, local governments will support the realization of the central government's aims. For instance, since protection of the right to the contracted management of farmers' land would not exert substantial influence on their own aims, local governments will support the central government to achieve its aim. In the same way, when local governments are able to control their economic growth rates and land development rates within some desirable range so as not to threaten the central government's policies of protection of cultivated land and farmers' status quo, the central government will also give its tacit agreement to local governments' actions and only makes certain requirements in terms of the scope and indicators of land use. If local governments go too far in land occupation, the central government will take measures to strengthen its control over the situation.

1.2.3. *The dynamic equilibrium between occupation and reclamation of cultivated land lacks assessment on scientific basis and standards*

First, no agreement has been reached in terms of people's understandings of the dynamic equilibrium of the total amount of cultivated land and there are generally four types of opinions: The equilibrium of total area of cultivated land in a particular region (including per capita cultivated land area); the dynamic equilibrium of cultivated land based on quality; the dynamic equilibrium of cultivated land based on grain productivity, and the dynamic equilibrium based on comprehensive indicators (the dynamic equilibrium of the comprehensive productivity of cultivated land resources), which is determined by multiple factors, such as the amount and quality of cultivated land, agricultural technology, capital and operation and management deployed, etc. Obviously, different understandings of equilibrium have different assessment standards and reflect their respective aims of equilibrium. For example, through treatment of saline soil and floods, medium-low yield fields could be transformed and productivity of cultivated land could be improved, and it would be entirely possible to realize the aim of food security even if the cultivated land amount decreases. This might satisfy the third or fourth kind of equilibrium although it cannot satisfy the first or second sort of equilibrium. Therefore, it is very difficult to set a standard to scientifically assess the dynamic equilibrium of the total amount of cultivated land when people's understandings and aims in achieving equilibrium are not the same.

Second, in practice, it is relatively easier to control the equilibrium of cultivated land area in a particular region and harder to control the equilibrium in terms of comprehensive cultivated land productivity, which includes land quality, food productivity, ecological and environmental factors.

1.2.4. Land use control does not take farmers' interests into account

Due to special protection of cultivated land, farmers' collective cannot change land use illegally even when they have the opportunity to increase their revenue by changing the use of agricultural land. In fact, in the course of implementation of the policy of cultivated land protection, the farmers' interests have been sacrificed without getting any compensation. However, under market mechanism, when the farmers and rural collective economic organizations cannot get reasonable compensation through the normal channels, they will attempt to gain this sort of benefits through gaming with the government, an example in case is the emergence of limited property right housing.

1.2.5. Self-supervision weakens the functions of land supervision

According to Article 3 of *The Land Administration of the People's Republic of China*, the people's governments at all levels should manage to make an overall plan for the use of land to strictly administer, protect and develop land resources and stop any illegal occupation of land. Article 6 of *The Regulations on The Protection of Basic Farmland* promulgated by the State Council also stipulates that "The village (township) people's governments shall be responsible for the work of administering the protection of basic farmland within their respective administrative areas". For one thing, local land and resources departments are under the control of local governments and cannot carry out land supervision functions independently. For another, local governments are the initiators of malpractice on land and it would be rather inefficient to ask them to supervise themselves.

During the past seven years since the implementation of the *new Land Administration Law*, i.e., from 1999 to 2005, over one million cases of illegal land behaviors have been found around the country and more than five million mu of land was involved, almost one million mu higher than 4.02 million mu, the total newly added construction land area of the country in 2004. This data also proved the defaults in the new *Land Administration Law*.

2. The Ineffectiveness of the Central Government's Campaign-style Land Law Enforcement

Apart from the continuous issuance and revisions of laws and policies, the campaign-style special investigation and prosecution actions launched by the central government and provincial governmental departments were also frequently reported. Although the amount and kinds of illegal cases are shown as achievements, they also raise such a question: Have illegal land behaviors been almost wiped out or is that just the tip of the iceberg?

2.1. *The land law enforcement campaigns by the central government are spearheaded at local governments*

One of the major tasks regarding land and resources management of the central government is to enhance land law enforcement and punish illegal land behaviors more severely. In 2003, the government carried out land market order rectification based on which the national land supervision system was established. From 2000 to 2006, 8,698 people were given disciplinary punishment by the party or the government and 1,221 people were investigated for criminal responsibilities. This, however, failed to curb illegal land behaviors and the central government was unsatisfied.

The central government carried out intensive special law enforcement campaigns in the year 2007, such as, the special campaign to crack down cases of violations of laws and regulations concerning land use by the Ministry of Supervision and the Ministry of Land and Resources, the special campaign to clean up the assignment of land use right by the Ministry of Supervision and the National Audit Office, the special campaign to address outstanding rural land problems by seven ministries and commissions, i.e., the Ministry of Agriculture, and the 100-day nationwide campaign for land law enforcement by the Ministry of Land and Resources. The strength and high intensity of land law enforcement were rather rare.

In October 2006, under the unified arrangement of the Central Committee of the CPC and the State Council, the Ministry of Supervision and the Ministry of Land and Resources issued the *Circular on Carrying out Special Campaign on the Investigation and Handling of Cases of Violations of Laws and Regulations Concerning Land Use.* Local supervision departments and land and resources departments at all levels concentrated time and energy to carry out a thorough clearing-up and investigation of newly added construction land occurring from July 2005 to September 2006 on a case-by-case basis in order to comb out clues regarding cases of illegal land use and to investigate and handle them. By the beginning of 2007, investigation carried out by various regions involved 32,872.84 hectares of land; 13,059 cases had been concluded in which 17,500.70 hectares of land was involved; among all the people given disciplinary punishment by the party or the government, there were two cadres at the prefectural or bureau level and 105 cadres at the county level; 879 people were transferred to judicial organs and 168 were given criminal punishment.

The Ministry of Supervision and the Ministry of Land and Resources issued the *Circular on Further Carrying out Special Campaigns for the Investigation and Handling of Cases of Violations of Laws and Regulations Concerning Land Use* in accordance with the requirements of the State Council and made arrangements on continuing to carry out special campaigns in July 2007. The Ministry of Land and Resources launched the 100-day nationwide campaign for land law enforcement in

September 2007. On October 12, 2007, the Ministry of Supervision and the Ministry of Land and Resources held the *Debriefing Meeting for Some Provinces (Regions, Cities) to Report the Progress of the Investigation and Handling of Cases Concerning Land-Related Violations.* Local people's governments, supervision departments and land and resources departments at all levels were requested to strengthen inspection and supervision and to impose more severe penalty so as to ensure that concrete results could be achieved in special campaigns. In pursuance of these requirements, local supervision departments and land and resources departments at all levels carried out "looking back" examinations of the special campaigns launched since October 2006, rectified newly added construction land occurring from October 2006 to July 2007, and a batch of new clues to violations of laws and regulations concerning land use were combed out. Meanwhile, efforts were generally intensified to investigate and prosecute illegal land-use cases.

Since November 2006, the Ministry of Supervision and the Ministry of Land and Resources formed joint inspection teams and carried out five rounds of inspections in 21 provinces, autonomous regions and municipalities, and 44 districts under them. Based on major problems of illegal land use found out during the inspection process and clues from public tip-offs, the joint inspection teams sent letters to supervise the handling of 50 cases of violations of laws and regulations in 13 provinces, autonomous regions and municipalities and urged relevant regions to investigate and handle such cases in strict accordance with laws and disciplines. By the end of 2007, the two ministries had examined and approved the investigation and handling opinions of 34 cases and 105 people were given disciplinary punishment from the party or the government, among which there were nine cadres at the prefectural (bureau) level and 47 cadres at the county level.

Another achievement of the land law enforcement campaigns was the discovery of various manifestations of violations of laws concerning land use.

2.1.1. *Breaking up the whole into parts*

According to *The Land Administration Law of the People's Republic of China, The Regulations for the Implementation of the Land Administration Law of the People's Republic of China*, and administrative laws and regulations formulated by the People's Congress at the provincial and prefectural levels, people's governments at all levels are endowed with the examination and approval power relating to land requisition of varying degrees, for example, the requisition of basic farmland or land outside the basic farmland exceeding 35 hectares, and other types of land exceeding 70 hectares shall be approved by the State Council. In order to circumvent the State Council, local governments cook up virtual partition of 1,000 or even 10,000 hectares of land occupied by large-scale development projects, and register the land occupation by one project under several different land certificates on the pretext that this is planning by stages, construction by stages and land requisition by stages.

2.1.2. Land requisition replaced by leasing

Land requisition, no matter how much land will be requisitioned, has to be approved by governments at corresponding levels, and corresponding compensations have to be paid to owners or users of requisitioned land. Therefore, it has become the first choice of many land speculators and enterprises or units attempting to occupy land to contract or rent collective land (including cultivated land, cropland) from villagers' committees, village groups and town or township governments for as long as 30, or even 40 or 50 years by paying merely corresponding land contracting or leasing fees to owners of collective land or cultivators of the original land. This sort of contracting or leasing is usually supported and protected by villagers' committees (mainly village leaders) and local governments (mainly at the township or town level), and the problem of illegal land occupation is rarely taken into account. As a result, the use of large amounts of farmland or cultivated land have been changed, even patches of factories or residences have being built on some of them. At present, "expropriation in lieu of leasing" is rather rampant among illegal land use cases and has become a common phenomenon. In order to circumvent the examination for approval procedures and to avoid payment of new construction land use fees, village groups in some regions conduct land leasing by themselves without going through the examination and approval procedures of conversion of agricultural land for other uses; some local governments step in to rent land from farmers and then lease these land to enterprises; some uses "reverse calculation method" to set land requisition compensation and resettlement fees so as to cover up the illegal land requisition compensation standard.

2.1.3. *Forced abandonment of cultivated land*

The Land Administration Law stipulates that "whereas occupation of land for construction purposes involves the conversion of agricultural land into land for construction purposes, the examination and approval procedures in this regard shall be required". In addition, in terms of laws, the examination and approval procedures of the conversion of agricultural land into land for construction purposes regulated by the state are very strict, and those of the requisition of basic farmland and cultivated land are even stricter. Besides, *The Land Administration Law* also stipulates that "land shall be used sparingly for non-agricultural construction purposes. Whereas wasteland can be used, no cultivated land should be occupied". Moreover, local governments have more examination and approval power on the occupation of wasteland. Therefore, in order to requisition rural land without going through the examination and approval by the State Council or higher-level governments, some local governments force farmers to let their farmland or cultivated land to go wasted (a small amount of compensation might be paid to farmers during the process, which is usually the compensation for green crops for two or three years) and then expropriate these abandoned farmland or cultivated land as wasteland. By this way, the local governments both achieve their aim of avoiding the examination and approval by higher-level governments and substantially reduce the compensation expenses to farmers. This is really a trick which kills two birds with one stone.

2.1.4. *Little requisition, much occupation*

Land requisition cannot be avoided in construction projects, such as, real estate development, industrial park construction, etc. If no land requisition procedures have been completed or no relevant approval has been obtained, it would be apparently impossible to cover it up from the public and even more impossible to cope with the supervision and inspection from the higher-level administrative departments. In such malpractices, land requisition procedure of 500 mu was completed for a project with 2,000 mu of land occupation, the procedure of 100 mu was completed for a project with 300 mu of land occupation, and the procedure of 30 mu was completed for a project with 50 mu of land occupation. In this way, relevant land requisition certificates or licenses could be shown to cope with inspections from the higher-level administrative departments or critical consumers who request relevant certificates or licenses when purchasing houses. In the ordinary course of events, this camouflage tactic of "little requisition, much occupation" is very effective because few people will closely inspect the tricks behind it. The worst situation will be that the requisition is said to be illegal excessive land use and it would be all right as long as the formalities are completed for the land occupied in excess.

2.1.5. *Requisitioning this while occupying that*

Namely, requisitioning a piece of land by using the land requisition approval for another piece of land.

2.1.6. *Planning modification*

In comparison with the examination and approval of land requisition for construction of houses, apartments or office buildings, that for industrial land, educational land or agricultural development land is usually much easier. Therefore, in order to seize and occupy land at cheap prices, individual real estate developers firstly purchase land from local governments in the name of industrial land, educational land, agricultural development land, etc., and then induce relevant governmental departments to revise the original contents of city planning, or even to allow them to proceed land use conversion procedures through active public relation activities with the government. For instance, some developers seize land in the name of low density residence development or even industrial projects, and then build villas or develop villas in golf courses and tourism resorts.

2.1.7. *Old buildings reconstruction*

If certain local governments do not intentionally requisition or occupy more land for non-agricultural construction, then large amounts of the already approved or completed projects could have been constructed by making use of the sites of old buildings and there is no need to occupy farmland or other land for constructions. For example, projects under such pretexts as New Rural Construction Projects, Old Village Renovation Projects, Dangerous and Old Village Houses Renovation Projects, etc., could absolutely conduct constructions by making use of the old house sites in rural areas and there is no need to occupy more farmland. As long as the Old-City Renovation Projects or the Dangerous and Old City Buildings Renovation Projects do not affect the overall city planning, there is also no need to construct new cities in other places or to construct new communities by requisitioning more land. Projects like "reallocation of governmental departments" may conceal deliberate intention to seize and occupy land because new pieces of land are requisitioned and allocated while the disposal of land and the old buildings at the old sites is not clearly explained.

2.1.8. *Investment promotion*

Local governments set up various development zones (including economic and technological development zones, hi-tech development zones, industrial parks, cultural and scientific parks, information technology parks, etc.) and real estate investment promotion projects. In addition, development zones are further divided into provincial-level, prefectural-level, county-level, and even town-level or township-level.

2.1.9. *Land occupation before approval — Joint development*

The Land Administration Law clearly states that "Local people's governments at all levels shall support rural collective economic organizations and farmers in their efforts toward development and operations or in starting up enterprises". In terms of starting up enterprises, the only capital owned by rural collective economic organizations or farmers is the small amount of land, which is also an attraction for many interest groups or foreign investors. As a result, the measly small amount of cultivated land is invested into enterprises as capital stock, and agricultural land is converted to construction land or joint development land. This is the so-called "joint development". Certainly, there are quite a number of local governments who have land in their hands (which might be temporarily requisitioned from farmers) and conduct joint development with foreign investors. Local governments usually have a high respect for foreign joint developers with great financial prowess, so no matter which piece of land (especially farmland) such investors favor, the local government officials normally will try their best to satisfy the requirements of these developers. This is done under the pretext of "necessity to promote investment". Naturally, the consequence of this sort of joint development is more occupation of farmland.

From the above-mentioned various manifestations of illegal land use, it can be seen that the governments at various levels are an indispensable link, and no illegal land use will occur if there are no violations of laws by the governments at various levels.

Zhang Xinbao, Director of the Law Enforcement and Supervision Department of the Ministry of Land and Resources, pointed out that the phenomenon of illegal land occupation by local governments was very serious, and almost all of the worst cases involved local governments or relevant officials. In terms of illegal land use area, cases involving illegal land use approval by local governments took up 80% of the total illegal land use area, while cases of malpractice involving citizens, individuals or enterprises took up only 20% of the total area.

2.2. Why the "feign compliant" local governments always win in the game?

Although the tax-sharing system and the real control over land use seem to be the causes for and the source of strength of local governments in their gaming with the central government over land use, the central government's tacit allowance for "rooms for violation" towards the local governments is actually the root cause of its repeated failures in preventing illegal land use behaviors.

2.2.1. The fiscal system gives local governments the motive force to game with the central government

The tax-sharing system reform was carried out in 1994 and the current fiscal revenue pattern in China is as follows: The central government takes up 50% of the total national fiscal revenue, provinces and cities take up 30%, while the rest of 20% go to the counties and townships, i.e., the ratio of five to three to two. According to the research of Wang Xiaoguang from the Academy of Macroeconomic Research of National Development and Reform Commission, from 1994 to 2005, the central government's revenue accounted for an average proportion of 52% of the total national fiscal revenue and the revenue of local governments at all levels accounted for 48%, while the central government's administrative power accounted for around 30% of the total in average and the administrative power of local governments at all levels accounted for around 70%. Though the financial power of local governments at all levels was gradually decreased, their administrative power was not reduced accordingly and a serious disproportionate situation occurred. This is the root structural cause of repeated failures in controlling local investment. At present, most counties and townships are deeply in debt, and they directly confronted with "the three rural problems", i.e., problems of agriculture, rural villages and problems of the farmers. The improper fiscal revenue arrangement is the major reason why the problems of agriculture, rural villages and problems of the farmers have hitherto remained unsolvable. The root reason of land problems also lies in this. The central government considers land as the lifeblood of the country while local governments regard land as their "money bag". The local governments' impulse comes from their benefits in the following three aspects. The first is the political achievement in one's official career. In order to promote investment at extremely low or even zero land price or to build vanity projects for which certain officials might be remembered, principals of local governments give implicit consent or even support to illegal land use behaviors. The second is fiscal revenue increase. It is commonly held that "land is the inexhaustible source of revenue". Land-transferring fees and other land related revenues have become the major source of local governments' finance. The third is rent-seeking benefits. Some officials in charge of the examination and approval of land use make use of their power to approve illegal land uses, thus resulting in rent-seeking corruptions. The status quo of land-related finance reflects the profit-seeking motive, which is induced by current land requisition system, of local governments; at the same time it highlights the defects in current financial and taxation systems.

2.2.2. The financial system deprives of financing channels for infrastructure construction in local regions

Among all the investment in urban infrastructure construction, budget funds account for less than 10%, land-transferring fees 20% to 30% and bank loans 60% to 70%. On what security do banks extend loans? The answer is land. Both governments and enterprises offer land as security for loans.

2.2.3. *Performance targets and the cadre evaluation and selection system induce local government officials to resort to land for fast returns*

Achievements in one's official career, GDP and GDP growth rate are three crucial factors in cadre selection. What does GDP growth rely on? It relies on investment, urbanization, infrastructure construction, expansion of city boundaries, large-scale civil engineering projects and construction of main roads and big squares. In certain economically underdeveloped areas, some officials wish to catch up with the developed regions and make their people live a more affluent life as fast as possible. However, in accelerating economic development, they would have to abide to objective laws governing economic development and to do things within their means and ability. If they rashly put up establishments and launch new projects without taking into account the reality and the people's actual capability, they will only waste both money and manpower, eventually, their achievements will become one belonging only to their generation, but burdens for many generations to come. In reality, some projects with serious pollution were launched in certain regions due to local government officials' anxiety to achieve quick success and get instant benefits, and some magnificent city squares were built in some underdeveloped regions without taking into account local people's actual need and capability.

2.2.4. *The land resources administration system lets local governments to possess the right to dominate over land usage*

State-owned land belongs to the state in name, but the actual right of dominion is to a large extent controlled by local governments. This provides convenience to local governments to indulge in land malpractices. In reality, local governments are also economic persons, to say nothing of the government officials who are also natural persons. Therefore, their instincts to seek profits and avoid losses and obvious rent-seeking impulse are sort of unavoidable. As one of the organizations of local governments, local land and resources administrative departments are under the control of local governments and the relevant leaders both in terms of organization and funds, and it is rather difficult to make them the loyal "watch dogs" of state land and resources. In fact, it is really rather awkward and unreasonable to ask local land and resources departments to play the double role of both the administrator and supervisor of state land and resources.

2.2.5. Defects in the law enforcement, supervision and penalty system reduce local governments' costs of breaking the law

The land law enforcement supervision and penalty system in China is rather weak, and punishment on large amounts of illegal land use cases is not adequate and fails to play the role of deterrence and warning. Although the land area involved in illegal land use cases increased by nearly 90% in comparison with that of 2005, judging from the most recent investigation and prosecution results released by relevant departments, the punishment was still too weak.

2.2.6. The central government tacitly allows rooms for corruption for local governments

Under the current arrangement of administrative power and financial power division between the central government and local governments, on the one hand, the central government strengthens its efforts in cracking down illegal land use behaviors when land disputes increase in number and affect social stability; on the other hand, the central government tacitly makes allowance for local governments because land profits reduce local governments' pressure in asking for finance help from the central government. They will form an alliance immediately and take a series of measures to block land rights defense actions when illegal land use behaviors provokes mass dissatisfaction. Certain tacit agreements exist between them.

It is estimated that local governments at all levels obtained a land net profit of over 1,420 billion RMB from farmers from 1987 to 2002 through cultivated land occupation for non-agricultural construction only.

2.2.7. There are blanks in enacted laws regarding crimes of village communist party secretary

Article 93 of *The Criminal Law of the People's Republic of China* has stipulated the applicable provisions for duty related crimes committed by village cadres. The Interpretation of the Standing Committee of the National People's Congress on Paragraph 2 of Article 93 of the Criminal Law of the People's Republic of China further defines the boundaries: "When a member of a villagers' committee or of any other village grassroots organization assists the people's government to exercise the following administration tasks, he shall be regarded as "any other state functionary performing public service according to the law" as mentioned in the second paragraph of Article 93 of the Criminal Law: (1) The administration of funds and materials for disaster relief, emergency rescue, flood prevention and control, special care for disabled servicemen and the families of revolutionary martyrs and servicemen, aid to the poor, relocating people and social relief; (2) The administration of funds and materials donated by the public for public welfare undertakings; (3) The administration of operation and management of land, as well as the management of house sites; (4) The administration of the compensations for land requisition; (5) Withholding taxes; (6) The tasks relating to family planning, permanent residence, etc.; (7) Assisting the people's government to exercise other administration tasks. Where any of the members of a villagers' committee or of any other village grassroots organization illegally embezzles any public money or property by taking advantage of his position during the course of performing any of the public services as mentioned in the preceding paragraph, if any crime is constituted, the crime of embezzlement as prescribed in Articles 382 and 383, crime of misappropriation of public money in Article 384 and crime of acceptance of bribes in Article 386 shall apply". In other words, a crime by taking advantage of duty by village cadres is constituted only when they assist the people's government to exercise the administration tasks.

However, there are still controversies in the theory circle regarding whether village communist party secretaries could be directly convicted if they hold no positions in villagers' committees. The first opinion is that village communist party secretaries could not be directly convicted because village grassroots organizations includes villagers' committees only and communist party organizations are not included. The second opinion is that village communist party secretaries could be directly convicted because the identity of village communist party secretary can be interpreted as "a members of villagers' committee or of any other village grassroots organization". Under the arrangement of present political system, besides villagers' committees, village communist party organizations also play an important role in village administration. In many places, the influence of the principals of village communist party organizations is no less than that of villagers' committees, and the opinions of village communist party organizations may play a crucial role in decision of many important issues within the villages. The third opinion is that village communist party secretaries could be directly convicted because they can be viewed as joint offenders.

3. Prospects of Land Property Rights Defense Activities by Farmers

Unlike the central government, the fundamental goal of land defense activities by farmers does not lie in the protection of the state's food security but in guaranteeing their basic source of income and in fair sharing of the added values of the land.

3.1. *Features of farmers' land property rights defense activities*

According to the investigation materials from March to June 2004 compiled by two research groups of The Institute of Chinese Academy of Social Sciences, i.e., Rural Development Research Institute's National Social Science Fund and the Major National Soft Science Research Projects, in the first half of 2004, 46,900 illegal land use cases in total were found all over the country, 33,900 illegal land use cases were investigated and prosecuted, 8.74 billion RMB or 59% of the total land acquisition compensation fees owed to the farmers were returned; 4,735 development zones of various kinds were retracted, accounting for 70% of the 6,741 development zones at the beginning of 2004, and the planned area was reduced from 37,500 to 13,400 square kilometers, a reduction of 24,100 square kilometers.

3.1.1. *Issues in dispute are rather definite*

The research group randomly selected 837 or 58% from the total of 1,434 petition letters which includes 1,325 from a media of the central government and 109 received by the research group, and made a quantitative analysis. It was found that the present rural land disputes mainly fall into two categories: Land acquisition and land occupation. From the contents of the petition letters, the nature of the problems can be classified as follows: 277 petition letters regarding illegal and forceful land acquisition, accounting for 33.1% of the total; 192 letters related to excessively low land compensation fees, making up 22.9%; 34 letters on poor resettlement arrangements for land-expropriated farmers, accounting for 4.1%; 185 letters regarding forceful occupation or privately dividing preserved collective land, accounting for 22.1%; 73 letters related to private land division or sales, making up 8.7%; 40 letters on forceful conversion of land use, accounting for 4.8%; and 36 letters regarding forceful withdraw of contracted land, accounting for 4.3%.

It is clear that the major problem reflected by the farmers belongs to disputes related to land acquisition, accounting for 60.1% of the total; out of which, illegal and forceful land acquisition accounted for 33.1% of the total.

3.1.2. Parties involved in conflicts have changed

The 837 petition letters were the main materials based on which the research team made an analysis of the plaintiffs and defendants involved in rural land disputes. The basic situation of the plaintiffs in the 837 petition letters regarding land problems is as follows: 629 joint letters by villagers, accounting for 75.1% of the total; 123 letters with stamps of villagers' groups, making up 14.7%; 34 letters with stamps of villagers' committees, taking 4.1%; 22 letters from other mass organizations, i.e., the rural senior citizen associations, accounting for 2.6%; 29 letters from individual villagers, making up 3.5%.

The basic situation of the defendants in the 837 petition letters regarding land problems is as follows: 102 letters regarding the prefecture-level governments and the administrative departments, accounting for 12.9% of the total; 221 letters related to county or district governments and the administrative departments, making up 26.4%; 217 letters regarding town or township governments and relevant leading cadres, accounting for 25.9%; 192 letters related to cadres of the village party branches or villagers' committees, accounting for 22.9%; 67 letters regarding development zones at various levels, making up 8%; 32 letters related to real estate developers and land acquisition enterprises, accounting for 3.8%.

The statistics indicated the following information: First, although joint letters by the villagers remained the prominent form in terms of plaintiffs, village organizations have become important plaintiffs mainly because their interests coincide with that of the farmers in forceful and illegal land acquisition disputes, so under the pressure of the villagers, they may become the main bodies in the struggle. It is not rare to see that all villagers, men and women, old and young, rallying together in certain real conflicts scenes. Second, city and county governments accounted for a high proportion of the defendants. However, in rural taxes and fees disputes, the main defendants were concentrated in the town/township and village levels, in which town/township governments were the major defendants, while city and county governments rarely became defendants. Last but not the least, enterprises and developers became defendants, which had never happened in previous rural taxes and fees disputes.

3.1.3. *The regional distribution of rural land conflicts has changed*

According to the statistics of the received phone calls and letters by a media of the central government from January 1–June 30, 2004, provinces with serious rural land problems include Shandong Province with 2,183 phone calls and letters, accounting for 14.3% of the total, Hebei Province with 1,756 phone calls and letters, accounting for 11.5%, Henan Province with 1,183 phone calls and letters, accounting for 7.7%, Heilongjiang Province with 1,023 phone calls and letters, accounting for 6.7%, Anhui Province with 930 phone calls and letters, accounting for 6.1% and Jiangsu Province with 926 phone calls and letters, accounting for 6.1%. The six provinces added up to 8,003 phone calls and letters, accounting for 52.3% of the total.

According to the statistical analysis of the 837 petition letters regarding land disputes, Zhejiang Province accounted for 12.8% with 107 letters, Hebei Province made up 9.3% with 78 letters, Shandong Province accounted for 8.6% with 72 letters, Jiangsu Province accounted for 8.1% with 68 letters, Guangdong Province made up 7.85% with 65 letters, Heilongjiang accounted for 6.7% with 56 letters, Anhui Province accounted for 6.4% with 53 letters, and other provinces added up to 46.7%.

In accordance with the statistics of the 87 land-related conflicts between the police and farmers occurred from January to June 2004, these incidents concentrated in the following regions: Zhejiang Province with seven cases,

Liaoning Province with seven cases, Jiangsu Province with six cases, Hebei Province with six cases, Shandong Province with five cases, Guangdong Province with five cases, and Gansu Province with five cases.

In view of the above-mentioned facts, the research group concluded that, unlike farmers' refusal to pay taxes and fees which mainly occurred in agricultural provinces in the central regions, the present rural land disputes concentrated in coastal developed regions and was most evident in Zhejiang, Shandong, Jiangsu, Hebei and Guangdong. Disputes in these regions were mostly related with illegal or forceful land acquisitions and the city and county governments were the major defendants accused by the farmers. In the central regions, such as, Anhui, Henan, Heilongjiang, etc., the major problem was the infringement on farmers' contracting rights, and organizations at the town, township, and village levels were the major defendants accused by farmers.

3.1.4. Conflict means have changed

In rural taxes and fees disputes, petitions, propaganda and collection resistance are the major struggle means. In rural land disputes, however, sit-down protests at the gate of the county or city government buildings or on the expropriated land, demonstrations, and even sit-down protests on the highway and railroad have become the main struggle ways. According to the statistics of the immediate causes for 87 land-related conflicts between the police and farmers occurred during January to July 2004, 48 cases were caused by farmers attempting to stop constructions on the expropriated or occupied land, accounting for 55.2% of the total; 31 case were caused by local governments ordering the police to stop farmers from petition, making up 35.6%; eight cases were caused by farmers' sit-down petitions at the gate of the city government buildings, on railroads and highways, or at communication hubs, accounting for 9.2%.

3.1.5. Conflicts have become fiercer, and conflicts between the police and farmers occurred from time to time

Since the central government has clear provisions on prohibiting the utilization of the police in solving conflicts regarding rural taxes, local government faces both political risks and farmers' resistance if they use the police to repress the farmer representatives in such conflicts. Therefore, except some individual incidents, conflicts between the police and farmers related to taxes and fees rarely occur. However, in defense activities related to property rights of land, huge interests are involved, disputes between the two parties were usually impossible to be conciliated, as such, it has become rather common for some local governments to order the police to repress farmers in defense of their rights.

3.1.6. *Intervention of External Force Changes the Nature of Conflicts*

Since land guarantees farmers' subsistence and involves huge economic interests, each incident of land disputes exerts significant social influences. As strugglers, farmers often turn to society for help through various channels, and some intellectual elites, mainly law experts, get involved in this for economic and political purposes or social influences. Among the above-mentioned 837 petition letters, 49 were written by lawyers or other law talents. Most notably, due to the intervention of intellectual elites, many incidents were detached from their original nature of land disputes and became political incidents. The case of over 20,000 immigrants in Tangshan City of Hebei Province removing the Secretary of Tangshan Municipal Committee of the CPC from the position of Representative of the National People's Congress because of insufficient land acquisition compensation; the case of over 10,000 immigrants in Qinhuangdao City passing motion to remove city-level officials for the same reason; the case of migrant workers in Ningde City of Fujian Province removing the Secretary of Ningde Municipal Committee of the CPC from the position of Representative of the National People's Congress because of cultivated land occupation; and the case of land-expropriated farmers in Fuzhou City of Fujian Province requesting to remove the mayor from the position of Representative of the National People's Congress, all of which occurred with the intervention of a non-governmental organization. This is distinctively different from rural taxes and fees disputes which rarely involve long-term intervention of external forces in an organized way.

3.2. Farmers' land property right defense incidents in 2006

3.2.1. The land right defense incident by the land-expropriated farmers in Dongjiao village, Putian city, Fujian province

In order to stop the start of a substation project, villagers of Dingcuo, Dongjiao Village, Xidu Town, Licheng District, Putian City, Fujian Province, had a serious conflict with the police sent by local governments and the employees of the unit using the land on November 11, 2006, and over ten villagers were injured.

3.2.2. The land right defense incident by the land-expropriated farmers in Foshan city, Guangdong province

Over 100 villages of Ditian Village, Nanzhuang Town, Chancheng District, Foshan City, Guangdong Province had a conflict with the local police on April 6, 2006 because of their dissatisfaction with the land acquisition compensation by Shanhu Electronics Factory and the arrest of the representative of the land right defense activities. A number of villagers were injured in the conflict.

3.2.3. The land right defense incident by the land-expropriated farmers in Sanzhou village, Shunde city, Guangdong province

In order the solve the problem of land acquisition, nearly 1,000 villagers of Sanzhou Village, Shunde City, Guangdong Province, besieged the 300 officials and overseas Chinese attending the opening ceremony of a large-scale modern granary on November 8, 2006, and resulted in a serious conflict between the police and the villagers.

3.2.4. The land right defense incident by the land-expropriated farmers in Taige village, Changzhou City, Jiangsu province

Hundreds of villagers of Taige Village, Xueyan Town, Changzhou City, Jiangsu Province blockaded the town government building on July 27 and

28, 2006 due to dissatisfaction with the illegal sales of land by local officials and cadres, resulting in traffic disruption and a bloody clash with the police. Certain villagers were injured and arrested.

3.2.5. *The land right defense incident by a farmer named he Guosheng in Yaosai village, Jiangyin city, Jiangsu province*

Due to disagreement with the housing demolition plan, the 43-years-old villager He Guosheng of Yaosai Village, Chengjiang Town, Jiangyin City, Jiangsu Province was killed at home in his fight against staff who sneaked into his house at night aiming to demolish his house by force on August 31st, 2006.

3.2.6. *The land right defense incident by the land-expropriated farmers in Xibaijiawang village, Taizhou city, Zhejiang province*

Villagers of Xibaijiawang Village, Jiaojiang District, Taizhou City, Zhejiang Province fought against the forceful land acquisition in their village by local governments, and a bloody clash between the farmers guarding their land and the police attempting to dispel the farmers by force on March 22nd, 2006.

3.2.7. *The land right defense incident by the land-expropriated farmers in Baiguo village, Zigong city, Sichuan province*

Because no economic compensation was paid for their acquisitioned land, farmers of the Baiguo Village, Hongqi Township, Zigong City, Sichuan Province occupied that piece of land, requested for a meeting with the local government, and had a conflict with the police attempting to dispel them by force on June 28, 2006. Several villagers were injured and arrested.

3.3. *Land property rights and the bottleneck of the political system*

Land privatization might be an important condition for farmers to defend their land rights, to obtain land profits and to improve land utilization efficiency. However, Property Rights of Land will not materialize if the present political system remains unchanged.

Judging from the result, the effect of cultivated land protection is limited although the state kept intensifying cultivated land protection; while the farmers have risen up and picked up the weapon of law and even sacrificed their lives to fight against cultivated land destruction behaviors. Then, where does the problem lie? In a country ruled by law, legislation is the result of negotiations and compromises among different interest groups. Land legislation in China, however, has become the one-sided act of the central government due to the absence of farmers' participation. Participation in legislation is not as simple as soliciting opinions from all sides through the media in that it requires all interested parties to enjoy certain seats in the legislation stratum. In China's election system, farmers, the majority of the population, however, are "widely represented by others". In other words, under the current political arrangement, where the vote power four rural residents is equivalent to one city dweller, farmers who have the most direct contact with land and have the strongest motive to protect cultivated land, are actually excluded from the negotiation table concerning allocation of land interests.

The state may, in the name of "public interest", take over rural collective land for its use, but "the public interest" in this case is largely limited to public facilities in cities which have nothing to do with farmers, whose rights to existence and property are excluded from the consideration of "public interest". As a result, land acquisition compensations are to be paid to rural collectives while the standard of compensation is decided by city people. Although the Constitution stipulates the right to "self-governance by the villagers'", the village communist party secretaries are in actual control of the financial power of villages while their appointments and removals do not come within the jurisdiction of laws and regulations.

Under such circumstances, at a time when the land value has appreciated tremendously, farmers could only passively accept rather than actively participate in the interest allocation. When the farmers turned to laws for protection, the judicial departments, which lack independence, could not be of any help, and some local higher people's courts even clearly stated that they do not accept and hear land acquisition dispute cases. When farmers united to fight against illegal acts, the liability exemption of "justifiable defense" was never applied to them, and they were deemed as the violators of social orders.

Chapter 3

Free Trade of Property Rights of Land is An Effective Land-Saving System — On Property Rights of Land Transaction System Reform

From the point of view of economics, whether a land system is effective or not does not lie in who owns the land but in whether the land could be traded. What is more important is that there are effective transactions of land among different groups of people, i.e., owners, operators or users, and among different uses, and through such dynamic processes optimized allocation of land is realized. However, several major influential opinions believe that transaction of property rights of land will bring about negative results. Among these opinions, three most important and most popular ones are: (1) Free trade of property rights of land will lead to the decrease of cultivated land area, and the reduction of grain crops will result in starvation, thus causing social unrest; (2) Free trade of property rights of land will lead to monopoly of land; (3) Free trade of property rights of land will cause a number of farmers to lose their land, the main source of their income and the guarantee to support their life after retirement.

This chapter is devoted to the discussion of the first opinion and certain corresponding solutions will be proposed.

3.1. Will Free Trade of Property Rights of Land Lead to the Decrease of Cultivated Land?

The so-called "decrease of cultivated land" in the language of economics refers to conversion of agricultural land to other uses. Since China is undergoing an urbanization process, we may assume that the decrease in cultivated land is mainly caused by conversion of agricultural land for city construction purposes. On the face of it, the productivity of land for city construction purposes is far higher than that of agricultural land, usually being scores or hundreds of times that of agricultural land. Once the free trade of land is permitted, would most of the land in our country be covered by cities and would the cultivated land area witness a substantial decrease? However, it should be emphasized that it is exactly such characteristics of the cities that will not make such a consequence happen.

Cities are political, economic, and cultural centers, and in particular trade hubs, precisely the reasons for their birth, and the productivity of cities comes from its concentration and agglomeration effects. Since industries within cities mostly involve trade, the so-called productivity of a city actually refers to the efficiency of trade. A "market" in the textbooks of economics refers to the market which is evenly distributed in a certain space. However, there are only 135 people per square kilometer even in China which has a huge population, and a person has to walk about several kilometers to be able to trade with 135 potential traders if they all live evenly on land, which is almost totally inefficient given the diversity of today's consumption. Cities depend on the agglomeration of people in realizing geographically the functions of markets. When a large number of people gather together in a small space, the transaction cost, mainly travel cost, will be substantially saved, and the market becomes a reality.

Apart from the functions of price formation and resource allocation mentioned in textbooks, the market has another important feature, namely, economy of scale, which means the larger the population is, the more efficient the market will be, and the lower the average transaction cost will be (Sheng Hong, 1992, pp. 154, 155). This feature is also manifested by network externality, which means the more people are participating in the market, the larger the transaction utility of each person will be, and the larger the aggregate transaction utility of the market as a whole will be. Agglomeration of cities, therefore, becomes a positive feedback process. In other words, more people are attracted to cities because there are already many people in cities. Two of the constraining factors to this are the rise in land prices and the crowdedness. For one thing, land is scarce in cities, the center of concentration and agglomeration, and the price of land will rise up because of the competition among large numbers of people for this sort of limited resources. For another, although the economies of scale and the network externality of markets have the special feature of transcending time and space and do not rely on space for existence, but the biological human bodies do occupy space. That's why crowdedness is a common trouble of all big cities in the world and can never be completely solved.

However, as the centers of concentration or agglomeration, cities by themselves imply that it is impossible for them to be evenly distributed on the territory of a country. In general, a city must be a form of organization which occupies little space. Even within the cities, as the distance from the center is elongated, the degree of concentration or agglomeration will be lowered, the economies of scale of markets will decrease, which means the saving in transaction cost will be reduced, and the aggregate transaction utility of the market will also be brought down. This causes the land prices to keep descending as the distance from the center of the city is elongated until the productivity of land for city equals to that of agricultural land (Fig. 3.1). This is the edge of land occupation by cities, farther from which the productivity of land for city construction purposes will be lower than that of farmland and no rational person will cross this boundary to develop cities.

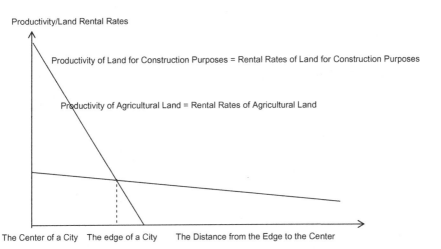

Figure 3.1. The Boundary Between Cities and Farmland.
Note: The oblique line on the left represents the productivity or rental rates of land for city construction purposes and the oblique line below represents the productivity or rental rates of agricultural land. The point of intersection of the two lines represents the boundary between cities and farmland.

Experience also shows that the above-mentioned "Law of Diminishing City Land Prices" is as irrefutable as "the Law of Diminishing Marginal Utility". In 2005, the average population density of the four core districts of function in Beijing, namely, Dongcheng, Xicheng, Chongwen and Xuanwu, was 22,210 people per square kilometer; that of the four function extended districts in Beijing, namely, Chaoyang, Fengtai, Shijingshan and Haidian, was 5,862 people per square kilometer; that of five inner suburban districts, i.e., Changping was 654 people per square kilometer; that of five outer suburban district, i.e., Yanqing was 198 people per square kilometer (Beijing Municipal Bureau of Statistics, 2006). We found out several examples from the information on land transfer for commercial and residential uses of 2007 from the Beijing Municipal Bureau of Land and Resources and roughly calculated the land prices of different districts in Beijing, which were listed in the following Table 3.1 in the order of the distance from the city center.

A more accurate conclusion will be drawn after a regression analysis of the prices of all land transfers which had already taken place. Moreover, it is assumed here that Beijing is a city with one single center while in fact Beijing has already developed in the direction of multi-centers. The basic trend of "diminishing city land prices", however, would be invariable.

On the other hand, the output value of agricultural land is also related to the distance from cities. Due to their direct service to cities, fast acquisition of demand information of cities, and high frequency of transactions with cities, the inner suburban districts enjoy lower average transaction costs and obtain higher incomes. By contrast, the outer suburban districts obtain lower

Table 3.1. Land Transfer Cases.

Land transfer cases	Districts	Price (RMB/ square meter)
Case I	Dongcheng District	10,281
Case II	Haidian District	6,901
Case III	Chaoyang District	5,012
Case IV	Changping District	3,318
Case V	Fengtai District	1,753
Case VI	Daxing District	490
Case VII	Yanqing District	150

Note: The above figures were calculated in accordance with materials from the Beijing Municipal Bureau of Land and Resources (http://www.bjgtj.gov.cn/).

incomes because of long distances from cities, low frequency of transactions and little knowledge about the demands of cities.

In the surrounding areas of cities, people acquire a rather high income through cultivation of flowers, nurseries and vegetables, production of meat, eggs and milk products, engagement in leisure and sightseeing, agriculture, and development of tourism and resort services. For example, the annual per mu revenue of greenhouse vegetables in Daxing District of Beijing was as high as 13,500 RMB (The Vegetable Office of the Daxing District People's Government, 2006); through multiplying this figure by 23, we get the approximate per mu land value of 310,500 RMB, or 466 RMB per square meter, which is very close to the land price of 469 RMB per square meter of Jufuyuan Community in Shiyi Village, Xihongmen Town, Daxing District which was formed through bid invitation, auction and listing. The average annual output value of the flower planting industry in the suburbs of Beijing was 13,900 RMB per mu (China Flower & Gardening News, 2007), and the land price was equivalent to 480 RMB per square meter; the land price had risen to nearly four million RMB per mu in Sansheng Towhship, Jinjiang District, Chengdu City due to flower planting and development of farmhouse enjoyment services (Gao Jikai and Pu wei, 2007), which was equivalent to 6,006 RMB per square meter. The land price determined by the output value of agricultural land is sufficient to constrain the expansion of cities.

Then, how large will land for city constructions expand to if free trade of land is permitted? In other words, how much farmland will be converted to city areas? This is a rather complicated question. However, we do not need an accurate calculation at present and a rough estimation will be enough. The core logic is that the agglomeration character enables cities to save more land than the countryside. According to information of the Ministry of Land and Resources, even the urban per capita land area which is considered to exceed the stipulated standards is merely 133 square meters, much lower than the rural per capita village land of 218 square meters. This means that 75 square meters will be saved whenever a villager is turned into a city dweller. In accordance with the population of 2004, if China's urbanization ratio reaches 70%,[1] or if 91,529 million people are

---

[1] As the scope of cities and towns defined in China is larger than that of many other countries (Li Qiang and Zhang Haihui, 2004), the urbanization ratio of 70% in China is used here to represent the urbanization ratio of 80% to 90% in other countries.

to live in cities, 122,000 square kilometers of land will need to be added for city construction purposes, which accounts for 1.27% of the territory. If we assume that the population would stop expanding, and 152.87 million people are to move into cities, then a city area of 20,000 square kilometers will need to be added, but a village land area of 33,000 square kilometers will be saved in the countryside at the same time. The net increase of land area will be 13,000 square kilometers.

Moreover, free trade of property rights of land will cause a substantial increase in land prices when agricultural land is converted for city construction purposes. The utilization efficiency of city land will be improved, and the urban per capital land area will decrease, thus saving a large amount of land. If the urban per capita land area is reduced from the current 133 square meters to 100 square meters and if the urbanization ratio stays at 70%, approximately 30,000 square kilometers will be saved all over the country.

Therefore, the cultivated land area will increase rather than decrease substantially once free trade of property rights of land is implemented both in terms of microeconomics mechanism and the macro aggregate figure. The reason is simple: the city in itself is a land-saving system.

3.2. The Current Land System, Land Polices and the Defaults

The current land system and policies, however, are based on the protection of cultivated land by using the government's coercive forces. The relevant systems include: (1) Protection of basic farmland; (2) Farmers are not allowed to change land usage by themselves; (3) In conversion of rural collective land to urban construction land, rural collective land is first acquisitioned by the government and then sold to developers after the primary development; (4) The area of rural house sites and residential land is limited, and the sale of house sites to people outside of the village is restricted.

Please refer to *The Land Administration Law of the People's Republic of China* and *The Law of the People's Republic of China on the Contracting of Rural Land* for details of these systems and policies. For example, the Land Administration Law stipulated "the dynamic balance of the total cultivated land area", "the balance between the occupation and supplement of cultivated land," "a strict control on the usages of land" and limitations on land use for building houses (Article 62). Although *The Law of the People's Republic of China on the Contracting of Rural Land* states that land contracting right can be transferred, it also clearly stipulates that "the agricultural use of land shall not be changed". Through local laws and regulations, the transaction of rural house sites is also restricted.

However, a thorough analysis shows that this sort of systems and policies aiming to protect cultivated land had an exactly opposite effect, leading to the excessive occupation of cultivated land and excessive land urbanization and bringing about other serious negative results.

First of all, since the use of "land ownership" of rural collectives has already been regulated, this sort of right has already been greatly weakened and its property value is therefore substantially depreciated. An asset might have many uses and its value is determined by its optimal use. If a specific use of this asset is regulated, the value of this asset will be evidently lowered by this regulation unless the regulated use is exactly the optimal use. At the least, due to urbanization and the expansion of cities, land in the suburbs of cities might become more valuable when it is converted from agricultural use to city use. In a market of free trade of land, the price of a piece of land in the suburb should stand between the agricultural use value and the city use value when this piece of land is converted from agricultural use to city use. However, since farmers' land ownership is restricted to agricultural use, it is impossible for their required land price to exceed the agricultural use value.

Second, although there are no explicit stipulations in laws, the "implementation regulations" regarding "the protection of cultivated land" and relevant policy regulations made it appear that in practice only the government has the right to convert agricultural land to city use, and the first step is the acquisition of rural collective land by the government. "Acquisition" is a coercive concept which indicates the deprivation of the rural collectives' right of negotiation on land transfer. When no fair negotiations are conducted, "land price", or the so-called land acquisition compensation, could never be fair. Theoretically, the party with coercive powers will eat up all the value increment between the city use value and the agricultural use value, which is manifested by the regulations of *The Land Administration Law* on land acquisition compensation. Article 47 of this law stipulates that "In requisitioning land, compensation should be made according to the original purposes (emphasized by the quoter) of the land requisitioned". The implied meaning is that the original land owners' right of negotiation on transactions of property rights of land is completely denied.

Third, due to defects of legislation in China, apart from the principles disadvantageous to farmers, the regulations on specific figures of compensation further slanted to directions against rural land owners. I once pointed out in my article entitled "How Much Should Be Paid to Land-expropriated Farmers as Compensation?" that "the true value of agricultural land should be the total discounted present value of all the future revenues of the land". "Future" here refers to for all time, "land revenue" refers to land rent, and "future revenue" refers to the endless revenue of land rent brought by land year after year. By now, there are no mature land tenancy markets in China and there is no balanced land rent rate for our reference. However, according to investigation of the periods when land could be freely traded and rented out, the market land rent rate was about 40% to 50% of the land output. We selected the mean value of 45% and the discount rate of 2%, and calculated that the agricultural land value in 50 years should be about 23 times higher than its average output value. If the discount rate is set at zero due to the consideration that land is the farmers' source of income handed down from generation to generation, the value of land will be even higher because the interests of future farmers, the descendants of present farmers, shall never be discounted (Sheng Hong, 2006). Article 47 of *The Land Administration Law*, however, regulated that "the land compensation fees shall be 6–10 times the average output value of the three years preceding the requisition of the cultivated land", which is substantially lower than the justified compensation level.

Moreover, the long-term agricultural policy implemented in China directly intervenes in farmers' production decisions and ensures the low prices of agricultural products. For one thing, farmers cannot decide which agricultural products to produce in accordance with the concrete conditions of land and local climate, and land resources are misplaced in the field of agriculture, thus bringing down the agricultural productivity. For another, the prices of agricultural products are forced down by the government's frequent price control on agriculture products and its monopoly on the purchase of agricultural products. Since the prices of factors are determined

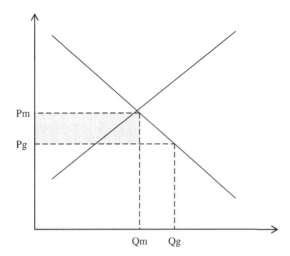

Figure 3.2. Demand for Land.

by the prices of products, the lowered prices of agricultural products lead to the underestimated prices of agricultural land.

Even the government has budget constraints. When the land price is far below the market equilibrium price, the demand for land will be greatly in excess of the equilibrium level (Fig. 3.2). As is shown in the figure, the land price determined by the market is *Pm*; at this price, land supply and demand equilibrium point is *Am*. When the government acquisitions land, the land price is forced down to *Pg* and the demand for land rises to *Qg*, which is far higher than the equilibrium level determined by the market. Therefore, the so-called systems and policies aiming to protect cultivated land actually resulted in more occupation of cultivated land.

At the back of the government are developers. As the government can requisition land at low prices and developers face the final product market, the requisition and sale price differential of land will be rather huge no matter through agreed transfer of land or land bid invitation, auction and listing. The price differential is either mostly obtained by the government (in the case of land bid invitation, auction and listing) or by developers (in the case of agreed transfer). Land acquisition at low prices and land sale at market prices become an important financial source of local governments and further enhance local governments' motive of requisition of rural collective land.

On the other hand, due to restrictions on transactions of rural house sites and the fact that urban governments view peasant workers as individuals rather than households, farmers entering cities either cannot move their families completely into cities, or have no means and no incentives to sell their houses and house sites in the countryside because the prices will be greatly lowered even if their houses and house sites could be traded as present laws and policies do not protect this sort of transactions. Therefore, the reduction of rural residential land caused by urbanization is greatly hindered, and urbanization in China looks like a one-way urban expansion process.

However, since the land is requisitioned at low prices, its utilization efficiency is also rather low. The misuse of land in cities of different regions was frequently reported. For example, "in Yinchuan City, whose population is less than 0.7 million, the annual demolition area has exceeded 2 million square meters since 2004. Seven large squares were built successively, and among them, the Datuanjie Square occupied over 0.3 million square meters, and the Xixia Square occupied 80 thousand square meters. However, few people visited these squares after their completion due to the small population" (Qin Yazhou and Wu Yong, 2007).

One piece of news from Xinhuanet.com pointed out that China's urban area was 399,000 square kilometers, the urban population was 338,050,000 in 2003, and the average density was 847 people per square kilometer (Xinhuanet.com, 2004). This indicates that each city dweller occupied 1,181 square meters which was higher than 1,150.4 square meters, the sum of the farmers' per capita cultivated land area all over the country (932.4 square meters) and the per capita residential land area (218 square meters). The agglomeration character of the city almost entirely disappeared, which is rather astonishing. Surely, this figure might be inaccurate due to different statistical sources. However, even the rural per capita land occupation of 133 square meters given by the Ministry of Land and Resources is still much higher than that of many market economy countries (82.4 square meters) and developing countries (83.3 square meters) (Dong Liming).[2]

The current land systems and policies lead to extremely low efficiency of land allocation, unfair distribution of wealth, and large numbers of social

---

[2]Quoted from Watching Out Four Phenomena of Land Wastage in Economic Daily News on 28 June 2006 (Xue Zhiwei, 2006).

conflicts. Around 74,000 recorded mass disturbances occurred in China in 2005, among which 16,312 cases involved land issues, accounting 20.7% of the total. Land is the Number 1 reason for conflicts (Yu Jianrong, 2006). Land disputes seriously threaten the harmony and stability of Chinese society.

3.3. Regulation for the Sake of Regulation

The low efficiency of land utilization, however, becomes the reason for further regulations by relevant governmental departments. In the beginning of 2007, the Standing Committee of the National People's Congress started examining *The Law of the People's Republic of China on Urban and Rural Planning* again. The draft law expanded the government's control over land from cities to the countryside because of the widespread misuse of land in China.

This is obviously an act which puts the cart before the horse. As pointed out in the above analysis, it is the excessive intervention in land allocation and the prohibition of free land transactions by the government that forced down land prices and resulted in excessive farmland occupation and the misplacement of land. This result then became the reason for further enhancement of the wrong system. Ironically, market economy countries which were used to prove the misuse of land in China actually implement the system of free trade of Property Rights of Land.

Some people proposed that China whose per capita land occupation was far smaller than that of the U.S. should learn from the U.S. which promulgated the *Farmland Protection Policy Act* in 1981 which clearly stipulated that the irreversible conversion of agricultural land to non-agricultural uses was restricted. However, after careful reading, you will find that this act mainly targeted the federal projects. Since the federal government and state governments occupy 47% of the total land area of U.S. and the federal government has strong fiscal strength, even the market price of land cannot effectively restrain the excessive use of land by the federal government. That's why the main content of this act was about the strict examination of federal projects which plan to change the use of agricultural land. Meanwhile, this act also declared that "the act does not authorize the Federal Government in any way to regulate the use of private or nonfederal land, or in any way, to affect the property rights of private owners."

Therefore, it is clear that the regulation on land use in the U.S. is the regulation on "Government Failure" rather than the remedy for "Market Failure". This act cautiously avoided harming present property rights of land system, indicating that land allocation in the U.S. still mainly relies on the market, based on which the government encourages people to preserve more farmland through subsidies, tax deductions, and trade of development rights.

China's higher population density and far smaller plain area than those of the U.S. do not indicate that the market system in China will not give effective reactions to such situations. As long as the market system functions well, it will result in more economized use of land in society with higher population density. High population density will exert influences in the following two aspects. First, in terms of consumption of agricultural products, the production of agricultural products with less land resources to meet the consumption demand of more people will lead to higher prices of agricultural products, and land prices will then become higher. Second, in terms of production of agricultural products, cultivation of less land with more labor force will result in higher land rent rates, and land prices will also become higher. This is to say, if the market exerts its functions, China's agricultural land prices will be higher than that of the U.S., which indicates a smaller space of cities and a closer distance from the edge of cities to the center under fixed urban population and population density. Moreover, the large population and the extremely small amount of rural per capita cultivated land in China will more easily attract people from the countryside to cities. Therefore, under the situation of high population density in China, the market system will strengthen the economized use of land. However, the urban per capita land occupation in China is roughly equal to that of the U.S.[3] which owns more land resources because the current land system in China destroys the market system and leads to serious government failure.

---

[3] According to Lu Dadao (2006), "in accordance with data released by the Ministry of Land and Resources of China, the comprehensive per capita land occupation of New York and its suburban area is merely 112 square meters. With its per capita cultivated land area being around ten times that of China, the U.S. is qualified to have a higher per capita land occupation. However, the U.S. did not do so. Though being quite high, the per capita land occupation in most large and medium cities of the U.S. still stands around 130 to 150 square meters".

Certainly, policies to strengthen regulations on land uses were made not only because of insufficient knowledge of the cause of excessive occupation of farmland but also because of the special feature of regulation, namely, regulation on land will also result in the formation of relevant interest groups. When rural collective land price (that is, land acquisition compensation level) is controlled, the regulation on land is quantity control and "rent" will occur at this moment. As is shown in Fig. 3.2, when land regulation departments attempt to reduce the land use amount from Qg to Qm, a rent rate of Pm-Pg will occur, and the total rent approximately equals to the area of the grey part in the figure. This rent will be divided among relevant government departments and developers. In order to obtain a piece of land, developers have incentives to bribe governmental officials. In reality, governmental departments of land and resources have already been called "one of the three areas with most serious corruption". According to the statistics of the Department of Law Enforcement and Supervision of the Ministry of Land and Resources in 2003, over 110,000 illegal land use cases were investigated and prosecuted, over 20,000 hectares of land were involved, 452 responsible persons were given administrative punishments, 771 people were given party discipline punishments and 168 people were given criminal punishments." High-ranking officials were caught for land issues in Beijing, Shanghai, Ningxia, Shenzhen, Qingdao, etc., and the total bribe or corruption amount was as high as 150 million RMB.[4] It is clear that land-related corruptions are not accidents occurring due to the lower moral standard of officials of land resources departments than that of officials of other departments but a systematic mistake of the current land system as a whole.

We can roughly conclude that China's current land systems that aim to protect cultivated land with the coercive forces of the government brings nothing but harm by being both inefficient and unfair, and creating opportunities for corruption of officials.

Attempts to impose more regulations upon current land regulations to solve the problems caused by regulations are actually like "quenching

---

[4] Such as, Yin Guoyuan, former deputy director of Shanghai Municipal Housing, Land and Resource Administration Bureau and former president of Shanghai Municipal Land Society. It was reported that "an unidentified property of over RMB 150 million yuan of Yin Guoyuan has been found out, including several apartments and large amounts of cash" (China Real Estate Business, 2007).

thirst with poison" and can never solve the basic problem of government failure. In other words, regulations will definitely bring about rent-seeking opportunities and the problem of who is to supervise the supervisors. A fundamental reform is to annul the present system which restricts the free trade of land.

3.4. Free Trade of Property Rights of Land

The property rights of land here include ownership of land, land contracting right, land use right, ownership of house sites, etc. They could be initial property rights, namely, rights regulated by the Constitution and laws (such as, ownership of rural collective land, etc.), property rights gained in accordance with legal procedures (such as, land contracting right, etc.), land property rights gained by reaching agreements through fair negotiations (for example, gaining land contracting right through land sub-contracting), and special land property rights gained through monetary transactions (such as, land use right, etc.), they could be permanent property rights or property rights with time limits.

Free trade includes the following aspects: (1) change in ownership of property rights of land; (2) change of land use; (3) decisions of the agreement content of transactions of property rights of land; and, especially, (4) decisions of transaction prices should not be intervened by any other individuals or organizations apart from the parties involved in the transaction as long as the transaction result does not cause any damage to the other people in the surrounding area of the transacted land.

Transactions of property rights of land could be the complete transfer of property rights of land, transactions of property rights of land divided into short intervals, such as, land lease which assigns the land use right for certain years, or transactions of property rights of land divided into several sorts of property rights, such as, the division property rights of land into "the right to the subsoil" and "the surface right" (the permanent tenancy right).

According to the above analysis, free trade of land will bring about three benefits: (1) restraint of the over-expansion of cities and economized use of land in the process of urbanization, and particularly the substantial reduction of costs in realization of this goal; (2) improvement of the income distribution structure in the process of land urbanization so that farmers will be able to enjoy the benefits of land value increment in the process of urbanization and the income distribution structure in China will be improved; (3) removal of the rent-seeking opportunities in land transactions so as to deal with governmental corruptions by "extracting the firewood from under the cauldron".

These three benefits are shown visually in Fig. 3.2. When we implement free trade of property rights of land, (1) the demand for land by cities will decline from Qg to Qm; (2) farmers will have more income when free trade of land is permitted than when trade of land is controlled, which roughly equals to the area of the grey part in the figure; (3) relevant governmental departments lose the rent-seeking opportunities of the area of the grey part at the most.

In fact, free trade of property rights of land brings about another important benefit, namely, it will set the market price of land. This price system gives people a correct signal so that land will be more effectively allocated both between cities and the countryside and within cities and the countryside. For example, different agricultural productions could be selected in different regions both in accordance with differences of geography, climate, and distance between cities. Suburban districts close to cities are more suitable for production of agricultural products with high transaction frequencies and products which are hard to preserve (such as, flowers, eggs, milk, meat, vegetables, etc.); districts far from cities can produce agricultural products with low transaction frequencies and products which are easy to preserve (such as, rice, wheat, etc.).

Particularly, free trade of property rights of land will conveniently convert rural house sites to city land and farmland and realize the economized use of land in the long urbanization process of China.

3.5. Institutional Conditions for Free Trade of Property Rights of Land

It should be admitted that an institutional framework of property rights of land has already been built in China which comprises the *Constitution, The Land Administration Law, The Law on Contracting of Rural Land*, etc. Under this legal framework, property rights of land in various forms have come into being, such as, property rights of rural land, ownership of house sites, land contracting right, land use right, etc. *The Property Law* which was adopted by the legislative body in 2007 further strengthened the process of documentation of various property rights so as to facilitate their transactions.

The existing laws and policies impose many restrictions but do not totally prohibit transactions of rural land. Restrictions on transactions of property rights of land include the following forms: (1) Restrictions on changes of land use, for example, only agricultural use is permitted; (2) Limitations on the identity of assignees, for example, transactions are allowed only among members of the same village; (3) Limitations on the types of tradable land, for example, only the use of four types of wasteland are allowed to be changed; (4) Regulations stipulating more complicated transaction procedures, for example, transfer of land contracting and operation right will have to solicit agreement from the party awarding the contract.

Under the constraints of these restrictions, transactions related to property rights of rural land are actually still going on. Even transactions changing land use and transferring property rights of land to people outside of the rural collective are being widely carried out in various regions of China, not to mention transfer of property rights of land without changing land use. Among those carrying out this sort of transactions are rural collectives, individual farmers, enterprises and even governments. For example: (1) Rural collectives directly construct production or commercial buildings (such as factories, office buildings, etc.) on rural collective land and then lease or sell these buildings to other enterprises or individuals; (2) Local governments in remote regions sell their urban construction land quotas to local governments of medium and large cities; (3) Local governments conduct long-term land lease transactions with rural collectives and convert these lands to city use; (4) Rural collectives directly construct residential buildings on their land and sell these buildings to urban residents and other people, which are the so-called "rural property right" or "limited property right" housing; (5) Individual farmers directly sell or lease their houses or house sites other people; (6) Rural collectives convert their land to shares and set up companies based on them, and make use of such companies to carry out industrial production and commercial project cooperation with other companies; and so on.

The existence of transactions of property rights of land indicates that such transactions will improve land allocation and accumulate large amounts of social wealth. However, since the present legal framework in China restricts and at least does not protect the transactions of property rights of land, the transactions results are doubted, transactions costs are high, and transactions themselves are repressed, thus hindering the improvement of allocation of land resources and the urbanization process in China.

In fact, the complete property right includes the transaction right. The limitation on transactions of property rights of land by present laws is actually an encroachment upon the completeness of property rights. Promotion of free trade of property rights of land will restore the completeness of property rights of land and improve the present system of property rights of land.

Therefore, in order to promote transactions of property rights of land, it is imperative to make amendments to present land-related legal framework. Most importantly, the false conception of so-called protection of cultivated land by the coercive force of the government must be completely denied. We suggest that:

(1) Article 17 (1) of *The Law on the Contracting of Rural Land* which stipulates that "Sustaining the agricultural purpose of use of the contracted land rather than utilizing such land for the purpose of non-agricultural construction" should be annulled.

(2) Article 33 (2) of *The Law on the Contracting of Rural Land* which stipulates that "the nature of the ownership of the contracted land and its use for agriculture shall not be altered" should be amended to "the nature of the ownership of the contracted land shall not be altered."

(3) "A strict control is to place on the turning of land for farm use to that for construction use to control the total amount of land for construction use and exercise a special protection on cultivated land" in Article 4 of *The Land Administration Law* should be amended to "A strict control is to be placed on the turning of state-owned land for farm use to that for construction use to control the total amount of land for construction use".

(4) Article 12 of *The Land Administration Law* which stipulates that "Changes of owners and usages of land, should go through the land alteration registration procedures" should be amended to "Changes of

owners of land should go through the land alteration registration procedures; changes of land usages should go through the filing procedure".

(5) "Peasants who have contracted land for operation are obliged to protect and use the land rationally according to the purposes agreed upon in the contracts" in Article 14 of *The Land Administration Law* should be amended to "Persons who have contracted land for operation are obliged to protect and use the land rationally".

(6) "The State protects the cultivated land and strictly controls the conversion of cultivated land into non-cultivated land" in Article 31 of *The Land Administration Law* should be amended to "The State protects the cultivated land and strictly controls the conversion of state-owned cultivated land into non-cultivated land".

(7) "The following cultivated land shall be demarcated as basic farmland protection areas and subject to stringent control according to the general plans for the utilization of land" in Article 34 of *The Land Administration Law* should be amended to "The following state-owned cultivated land shall be demarcated as basic farmland protection areas and subject to stringent control according to the general plans for the utilization of land".

(8) "Any unit or individual that need land for construction purposes should apply for the use of land owned by the State according to law, except land owned by peasant collectives used by collective economic organizations for construction of township commercial buildings or building houses for villagers or land owned by peasant collectives approved according to law for use in building public facilities or public welfare facilities of townships (towns)" in Article 43 of *The Land Administration Law* should be amended to "Any unit or individual that need land for construction purposes should apply for the use of land owned by the State according to law, except land owned by peasant collectives used by collective economic organizations for building township enterprises or building houses for villagers or land owned by peasant collectives approved according to law for use in building public facilities or public welfare facilities of townships (towns)".

(9) "In requisitioning land, compensation should be made according to the original purposes of the land requisitioned" in Article 47 of *The Land Administration Law* should be amended to "In requisitioning land,

compensation should be made according to the market value of the land requisitioned".

"Compensation fees for land requisitioned include land compensation fees, resettlement fees and compensation for attachments to or green crops on the land. The land compensation fees shall be 6–10 times the average output value of the three years preceding the requisition of the cultivated land. The resettlement fee shall be calculated according to the number of agricultural population to be resettled. The number of agricultural population to be resettled shall be calculated by dividing the amount of cultivated land requisitioned by the per capital land occupied of the unit whose land is requisitioned. The resettlement fees for each agricultural person to be resettled shall be four–six times the average annual output value of the three years preceding the requisition of the cultivated land. But the maximum resettlement fee per hectare of land requisitioned shall not exceed 15 times of the average annual output value of the three years prior to the requisition" in the same article should be amended to "Compensation fees for land requisitioned including land compensation fees, resettlement fees and compensation for attachments to or green crops on the land are to be viewed as the bottom line of compensation. The land compensation fees should be calculated in accordance with the discounted present value of all the future revenues of the land". "Future" here refers to for all time, land revenue refers to the reasonable land rent rates determined by the market, the discount rate should be the social discount rate which is lower than the market discount rate. The resettlement fees should be calculated in accordance with the average unemployment insurance level of urban workers in the neighboring cities.

The statement "In requisitioning vegetable fields in suburban areas, the units using the land should pay new vegetable field development and construction fund" in this article should be annulled.

The statement "But the combined total of land compensation fees and resettlement fees shall not exceed 30 times the average output value of the three years prior to the requisition" in this article should be annulled.

The core idea of these law amendment suggestions is that free trade of property rights of non-state-owned land should be protected and the existing provisions on protection of state-owned farmland in present laws should be preserved. However, under the revised legal system, if the government wants to preserve more farmland from the non-state-owned land, it shall have to make use of market-oriented policies and measures, for example, giving subsidies to rural collectives or individuals who preserve cultivated land, or learn from the experiences of other countries, allow land development right transactions, etc. The significance of these changes lies in the following aspects: (1) Operating merely on the marginal farmland, thereby substantially reducing the operational costs; (2) The government rather than the farmers bears the costs of farmland protection so that the income distribution in the process of land urbanization will head towards the direction of benefiting farmers.

3.6. Conclusion

The core idea of this chapter is that in most cases the rational choices by the market mechanism and individuals are far more powerful and efficient than governmental interventions in terms of realization of economized use of scarce resources. China's market-oriented reform in the past 30 years has eloquently proved this point. Surprisingly, the theory based on which the planned-economy was built continues to dominate the field of land resources allocation in China until now. The time has come for us to apply the simple truths revealed in China's reform and opening-up in the reform of land system.

Chapter 4

Literature Review on Food Security

In economic researches, food security is an extension of the internal balance between food supply and demand in a country or region, which is usually measured with the net food import (or food self-sufficiency rate) after the analysis of the balance between supply and demand. Therefore, domestic and international researches on food security are actually mostly application researches which assess the domestic food balance and security status through empirical investigations of domestic production and demand and international market status. This chapter attempts to sort out certain basic issues on food security and to carry out a literature review regarding the connotation and evolvement of food security, root causes threatening food security and relevant solutions, and the measurement indicators of food security, etc.

1. The Connotation and Evolvement of Food Security

1.1. Background of the concept

During the several thousand years of evolution of human civilization, especially during the time before industrial revolution, agriculture was the basis for the sustenance of human beings most of the time. It was, therefore, impossible for the food problem to be neglected by the people. However, it is only in modern society when human beings are able to get rid of the long-term hunger and starvation (Fogel, 1991, quoted by C. Peter Timmer, 2000.)

It was the worldwide food crisis[1] during 1972–1974 which drew international attention to the food problem and pursued the guarantee of food security as a common goal. This food crisis greatly promoted researches and discussions on food security. In November 1974, the Food and Agriculture Organization (FAO) organized the World Food Conference in which *Universal Declaration on the Eradication of Hunger and Malnutrition* was adopted. Meanwhile, the *International Undertaking of World Food Security* was adopted by the board of directors of the FAO. Since then, the issue of food security has been discussed.

1.2. The connotation and evolvement of the concept

Food security has many definitions and its connotation develops continuously as world food situation changes. The following passages introduce the widely accepted definition of food security and the issues of concern in the international community.

Food security, according to the definition proposed in the World Food Conference held by the FAO in 1974, refers to basic rights of life enjoyed by human beings, namely, "to assure access by all people at any place to enough food for existence and a healthy life". At the conference, it was also determined that the stocks-utilization ratio of 17%–18% was the minimum level necessary to safeguard world food security, in which revolving stock comprises 12% and reserve 5%–6%.

In April 1983, Edouard Saouma,[2] the former Director-General of FAO, proposed a new definition of food security and enriched its meaning, namely, "access by all people at all times to enough food for an active, healthy life". This definition includes three basic objectives: Guarantee of production of

---

[1]In the early 1970s, adverse weather in several consecutive years led to worldwide cereal output reduction. In addition, the Soviet Union changed its past practice of slaughtering livestock when domestic cereal output reduction occurred, and purchased large quantities of grain from the international cereal market. As a result, the severest food crisis after World War II came about. From 1971 to 1974, the international wheat price went up by 103% and the international corn price rose by 58%. Meanwhile, world food reserve fell from 77 days to 33 days.

[2]Edouard V. Saouma, the former Director-General of FAO from 1976 to 1993. Source: the official website of FAO.

enough food, stabilization of food supply to the fullest extent, and ensuring that all the people who need food have access to food.

A Plan of Action was adopted during the second World Food Summit in 1996 and 186 countries promised to "implement policies aimed at eradicating poverty and inequality and improving physical and economic access by all, at all times, to sufficient, nutritionally adequate and safe food and its effective utilization". The three fundamental elements[3] of food security widely accepted by FAO and USAID were also adopted by many scholars, namely, food availability which refers to sufficient quantities of food supply from domestic production and imports; food access which refers to the households' and individuals' abilities to procure food of adequate quantity and nutrition; food utilization which includes emphasis on health and health services, knowledge on nutrition and child health care, correct grain processing and storage technology (USAID Policy Determination, 1992, United States Department of Agriculture, 1996, P. 2). Frank Riely *et al.* (1999) gave a framework for the three fundamental elements (Fig. 4.1).

Figure 4.2 shows a framework of the connotations of food security. It is a longitudinal and complete concept of food security from the global level, to the national level, and then to the household and individual levels.

It can be seen from the above figure that food security in the broad sense includes both access to food at the global, national, household and individual levels and nutrition security. Food security at the global and national levels involves all links of food production, namely, production, storage, trade, etc. Food security at the household and individual levels is closely related to income levels and the relationship between food security and poverty is involved here. Therefore, we have to investigate the reasons threatening food security from these aspects.

Besides, there is another widely accepted classification of food security, namely, chronic food insecurity and transitory food insecurity. Chronic food insecurity refers to the long-term insufficient food supply and malnutrition due to the inability of a country or a household to produce or purchase enough food. Therefore, the reason for the chronic food insecurity in a

---

[3]FAO often lists stability as the fourth fundamental element of food security. Please see the Assessment of the World Food Security Situation, the Report of the 33rd Session of the Committee on World Food Security.

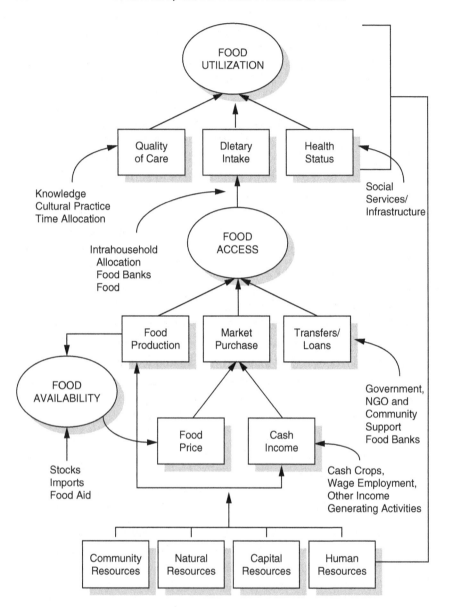

Figure 4.1. The Conceptual Framework of Food Security.

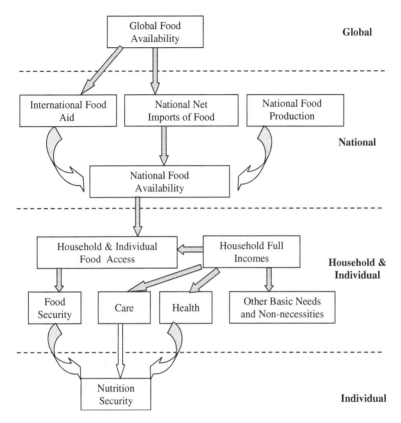

Figure 4.2. The Framework of the Connotations of Food Security.
Source: quoted from Lisa C. Smith *et al*. The geography and causes of food insecurity in developing countries, p. 201. The author made certain amendments to the content of the national level.

country is due to low production capacity and the rapid population expansion, which could be explained by the Malthusian Theory of Population. Similarly, the reason for the chronic food insecurity of a household is the long-term low income which is not enough to purchase sufficient food. The transitory food insecurity is most common in modern society and the causes include fluctuations in production, prices and income, crop failure caused by natural disasters, e.g., drought and the snow disaster in south China in the beginning of 2008, political systems against market economy, e.g., the political system of North Korea, political upheavals, e.g., the genocide in Africa.

2. What are the Root Causes Threatening Food Security?

The key to ascertain the root causes threatening food security is to examine the connotations of food security. Food security involves various aspects, generally speaking, the following aspects[4] are involved.

2.1. *Fluctuations in production and prices: Rising food prices due to short food supply caused by the decline of food output at the global and national levels*

As the worldwide food crisis in the early 1970s is the direct cause of people's concern for food security issues, food supply and food price fluctuations at the global and national levels were issues most concerned by the academic circle at that time.

There were abundant researches on food security in the 1970s. For example, the 13 theses collected by a conference jointly held by the International Food Policy Research Institute (IFPRI) and International Maize and Wheat Improvement Center (CIMMYT) in 1978 almost covered all the main ideas on food security at that time (Alex F. McCalla, 1982), such as, food security was mainly threatened by fluctuations in food production caused by climate changes and the consequent fluctuations in food prices and supply. In addition, international solutions to food security problems were put forward, included commodity agreement,[5] insurance

---

[4] Another classification from FAO divides food crisis into insufficient food supply, blocked access to food, and serious local problems. The three types are not mutually exclusive. First, the large gap of aggregate food output or food supply occurs due to crop failure, natural disasters, discontinued imports, disturbed distribution, excessive loss after harvest, and other supply obstacles. Second, majority of the population in countries who faced difficulties in acquiring food in many regions were recognized as incapable of purchasing food from local market due to extremely low incomes, extortionate food prices, or no circulation of food in domestic market. Third, serious local food insecurity occurs in countries with an influx of immigrants, gathering of domestic homelessness and migration of people, crop failure and serious poverty in certain regions. Source: FAO website http://www.fao.org/docrep/010/j9940c/j9940c09.htm.

[5] Agreements reached by producers and consumers to adjust production and prices of certain raw material commodities. Producers might agree to limit export through quotas, stocks or production reduction. The actions of the parties to this sort of agreement are sometimes

schemes, stocks, food aid and compensatory financing facilities.[6] Policy measures in the following four aspects were proposed: production, storage, consumption, and trade. Anthony H. Chisholm and Rodney Tyers (1982) believed that food insecurity at the national and global levels was actually the result of short-term fluctuations in food production and price. They also proposed two solutions to relieve food insecurity: food stocks and trade.

Figure 4.3 shows some important data about food output and price in the world, such as, the time-series changes in the aggregate cereal output and the per capita output, and food price changes represented by rice and wheat (the nominal price and the real price). It can be seen that the real prices of rice and wheat gradually declined; world food output showed a rising trend since the 1950s, and the per capita food output leveled off due to the expansion of world population.

It was generally believed that developing and underdeveloped countries were the most adversely affected by the rise in international food prices, and the per capita food consumption would decrease *vis-à-vis* the rise in international food prices, resulting in food security problems. However, Robert Paarlberg (2000) showed that developing or underdeveloped countries have limited dependence on world food market, and, between the international food prices and the per capita food consumption, there did not exist a simple inverse relationship. Statistics showed that while the international food prices started to rise around 1995 and the wheat export price of the U.S. went up from 157 U.S. dollars/Ton in the year 1994–1995 to 216 U.S. dollars/Ton in the year 1995–1996, the per capita cereal consumption of developing countries did not show any declining trends, instead, it rose from 170 kilograms in the year 1994–1995 to 171 kilograms in the year 1995–1996, and even reached 172 kilograms in the year 1996–1997.

There is no simple linear relation between the world food prices and the food security of developing and underdeveloped countries. It was the rising food prices which placed food security on the research agenda. However, the answer remains complicated as to how much developing countries are

---

triggered when market prices reach certain trigger price levels. This sort of agreement usually only exerts limited short-term effects.

[6]Compensatory Financing Facility (CFF) was set up in 1963 and then replaced by Compensatory & Contingenting Facility (CCFF). CFF or CCFF was established so that member states of IMF facing difficulties in international balance of payments, caused by decline in exports or increases in cereal import bills, may apply for loans other than ordinary loans.

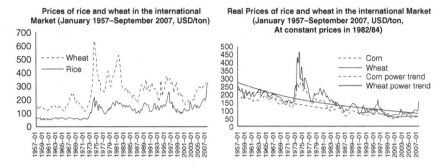

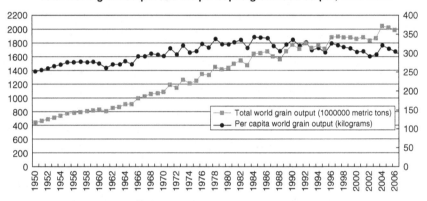

Figure 4.3. Total World Grain Output and the Per Capita World Grain Output, 1950–2006.
Source: http://www.earth-policy.org/Indicators/Grain/2006_data.htm.
Note: the above two figures were quoted from *The Supply of and the Demand for Grains in China and the Trend of Prices (1980–2007) — A Study on Grain Price Fluctuations, Macro Stability and Grain Security* by Lu Feng, and the data of the figure below was quoted from the data of United States Department of Agriculture from Earth Policy Institute and adapted by the author.

affected by the world food market. Research perspectives regarding this aspect include world food balance and food security, globalization and poverty, food reserve.

2.2. Poverty problem: Access to food/nutrition security at the national, household and individual levels

Strong and direct relationship exist among agricultural productivity, hunger and poverty. Three-quarters of the poverty stricken population live in rural areas and earn their livelihood by agriculture.

Food security and the problem of poverty are closely related. Food insecurity and hunger are two potential indicators of poverty (Helen H. Jensen, 2002). Although food supply is important for some countries, poverty is the root cause of food security problems in a large number of developing countries because poverty makes it difficult for the people in these countries to acquire food (Robert W. Herdt, 1984; Luther Tweeten, 1999; Handy Williamson, 2001). There were certain researches on income and poverty in the 1970s who believed that in some developing countries food insecurity was the result of unequal income distribution and lack of food allocation rather than the result of insufficient food production and foreign exchange reserve. Access to food and poverty became the theme of the second World Food Summit in 1996.

Since the 1990s, poverty has become the greatest obstacle to food security of people in developing countries. Therefore, the solution to food insecurity and the solution to poverty are two inseparable aspects where the solution to poverty is a sufficient condition for the solution to food insecurity. Lisa C. Smith *et al.* (2000) believed that the key to the solution to food insecurity is to improve people's capability to access to food through the poverty alleviation. They also proposed three ways to alleviate poverty. First, it should be ensured that the poor are able to share the fruits of economic growth, such as, information, infrastructure, financial resources, etc. Second, capital investment on human resources should be allocated to the poor so as to improve their health and educational level. Third, social capital of the poor should be enhanced so as to strengthen their ability in mutual help and trust.

Luther Tweeten (1999) pointed out that both long-term and temporary food insecurity traced to poverty and that poverty must be addressed by sustainable economic development. The author believed that the reasons for food insecurity in modern society lie in unfair political institutions and that food security was related to both agricultural development and economic development as a whole. The reasons why people are unable to access food include: (1) lack of purchasing power, and under such a condition, economic productivity will decline; (2) high food prices due to lack of free trade of food or insufficient food stocks, lack of transportation means etc. The primary solution to food security is improvements in the productivity and real income of the food-insecure group so as to improve their capability to access food.

The solution to poverty is the premise for the solution to food insecurity, and the poverty problem cannot be viewed in isolation. A number of scholars considered poverty rate as an important indicator of food security. At the national level, the development of a country's economy and the improvement of productivity enable the people to share the fruits of economic growth and for the poverty rate to decline so that the food security level of the country will be greatly enhanced.

2.3. Income distribution[7] — Amartya Sen's entitlement theory

The traditional explanation of food insecurity (e.g., famine) is insufficiency of food supply — Food Availability Decline (FAD) which is also the idea of Malthus. According to Malthus, human population was growing geometrically while food supply was growing arithmetically, and famine would occur when the growth of the food supply fell behind the growth of the human population. However, the decrease in food supply is not the only reason for food insecurity, and the uneven sharing of food among different income classes is another important factor. Amartya Sen put forward this idea or the Entitlement Approach in his *Poverty and Famines: An Essay on Entitlement and Deprivation*. According to Sen, entitlement is the most important one among all legal means to acquire and control food, famine occurs due to food entitlement failure of some people, and a person's ability to command food — indeed, to command any commodity he wishes to acquire or retain — depends on the entitlement relations that govern possession and use in that society. Sen interpreted famine from two aspects: direct entitlement failure and trade entitlement failure. The former is acquired through direct production while the latter is acquired through market trade. Therefore, in Sen's Entitlement Approach, direct entitlement failure refers to decline in food production which was mentioned in previous

---

[7]It was Shlomo Reutlinger and Marcelo Selowsky who first connected income distribution with hunger. They roughly estimated a relation between calorie consumption and income; then they estimated the calorie consumption for different income classes. On this basis, they could estimate the number of people whose supply of calories fell below some critical level. This approach gets at the underlying determinants of hunger.

paragraphs as the first reason of food insecurity, and trade entitlement failure is a concept related to income distribution.

2.4. *Other factors*

Other factors include natural disasters, unrests and political forms of government which are against the market principle. Robert Paarlberg (2000) believed unrests had become the most dangerous killer increasingly threatening food security among the three factors causing short-term food insecurity (unrests, disasters and political form of government which are against the market principle). In 1996, FAO listed 14 African countries South of the Sahara Desert as regions with food emergencies, among which 10 countries suffered food shortages caused by civil unrests. In the same year, the Office of Foreign Disaster Assistance of United States Agency for International Development listed 23 countries as regions with complex humanitarian emergencies, among which 19 countries faced food insecurity (of these, in 17 countries, food insecurity was caused by civil unrests).

The most recent assessment by FAO showed that food emergencies existed in 34 countries all over the world, 26 of which located in Africa and the remaining countries are located in Asia with a few exceptions. Table 4.1 cites the regional distribution of countries with serious food insecurity in the world and their causes. It can be seen from the figure that two main causes threatening food security are the present or most recent civil strife or conflicts and unfavorable climate conditions.

3. Indicators of Food Security or Insecurity

Indicators frequently used to assess food security include food production, income, aggregate expenditure, grain expenditure, the ratio of food expenditure to grain expenditure, food reserve, energy (calorie) consumption, nutrition situation, etc.

Food and Agriculture Organization of the United Nations (FAO), World Bank, International Food Policy Research Institute (IFPRI), and U.S. Department of Agriculture (USDA) generally use world food prices and food stocks as indicators of food security.

Lisa C. Smith *et al.* (2000) proposed three indicators of food security and nutrition security. The first indicator was daily per capita dietary

Table 4.1. Regional Distribution of Food Emergencies in the World and the Causes.

Regions with food emergencies	Representative countries	Causes for food security or emergencies	
Western and Central Africa	Chad	Deteriorating food security situation	Population displacement
	Mauritania and Niger	Reduced production	
	Cote d'Ivoire, Guinea, Liberia, Sierra Leone and the Central African Republic	Adverse weather	Refugees as a result of civil conflicts
Eastern Africa	Somalia	Unfavorable climate conditions(floods)	
	Eritrea	Unfavorable climate conditions(draught)	Civil conflicts
	Ethiopia Kenya	Large numbers of chronically food insecure people Unfavorable climate conditions(heavy rains and floods)	
	Sudan		Civil conflict RVF
	The United Republic of Tanzania	Unfavorable climate conditions(heavy rains and floods)	
	Uganda	A poor cropping season	Civil conflict

(Continued)

Table 4.1. (*Continued*)

Regions with food emergencies	Representative countries	Causes for food security or emergencies
Southern Africa	Zimbabwe	Economic crisis
	Angola Democratic Republic of Congo	Large numbers of vulnerable people Civil strife
Far East Asia	/	Political instability Unfavorable climate conditions
Near East	Iraq	Conflicts and insecurity
Asian CIS	Armenia	Unfavorable climate conditions(draught)
Caribbean	Haiti	Economic crisis

Source: Assessment of World Food Security Situation, Report of the 33 Session of the Committee on World Food Security, 7–10 May 2007, Rome. Adapted by the author.

energy balance (DEB). To investigate DEB, two other concepts need to be investigated first, namely, daily per capita dietary energy supply (DES) and daily per capita dietary energy requirement (DER). If the difference between DES and DER is greater than zero, macro food security is achieved. DEB is an indicator used to measure food availability at the national level. The second indicator is absolute poverty line, that is, the percentage of the population with an income of less than one dollar per day. This indicator is used to measure people's ability to food access. The third indicator is child malnutrition rate which is used to measure nutrition security, specifically the percentage of children under five years old who are underweight.

Indicators of food security include food supply (calorie supply per capita and protein supply per capita) and the percentage of healthy children under five years old (J. Craig Jenkins and Stephen J. Scanlan, 2001).

Besides, poverty is also used to assess food security level, and the most widely used method is Engle's Ratio Method to measure poverty line. The Engle Coefficient is the proportion of expense on food to the consumption expense which falls as family income rises. In other words, the bigger the Engle Coefficient is, the poorer a country is. The American scholar Orshansky studied the Engle curve of the consumption structure of American families and proposed that a family which spends higher than 30% of its budget on food expenditure is poor. The Absolute Poverty Line is a generally used indicator of poverty.

The FAO statistical database (FAOSTAT) gives a very detailed indicator of food security and releases some data at regular intervals which we can use for reference (see Fig. 4.4).

4. Solutions to Food Security

4.1. *State policies or interventions*

Food policies do not always measure the three parts of food security easily, namely, availability, acquisition and utilization. Sufficient food production and supply at the national level is important, but not the only requirement of food security. Policies also have to take into account the importance of subsistence and the importance of fair acquisition of resources so as to guarantee sufficient acquisition of food. Plans need to be made to ensure that economic growth and trade policies support policies on poverty

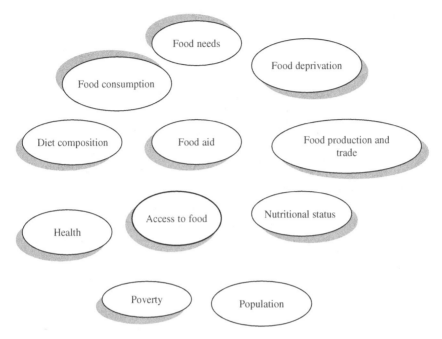

Figure 4.4. The Indicators of Food Security in FAOSTAT.

alleviation and food entitlements. When people cannot support themselves, plans need to be evolved to solve the problem of the inability of the poor to acquire sufficient food. For example, until recently, the food security policies of Southern Africa and India might have been overemphasizing food availability and the increase of food output so as to ensure self-sufficiency at the national level without paying enough attention to other factors hindering people's access to food.

Since the 1960s, the food policies of India have been aiming at promoting food and agricultural production and ensuring that the poor have access to the minimum food supply through public distribution system. Its agricultural development strategy placed much emphasis on providing subsidies to electricity, water and fertilizers inputs, etc. Private capital increased rapidly, especially in the green revolution areas where agriculture developed quickly, whereas investment in village roads, irrigation and electric power remained insufficient. The national agricultural policy adopted by the Indian government (2000) and the relevant plans implemented by her, attempted to rectify these deficiencies — different regional strategies were drafted, taking into consideration climates and natural resources conditions; more efficient mechanism was established for the implementation of decentralized food purchase policies.

The current food security policy implemented by South Africa is different from the self-sufficient policies implemented by its former Apartheid regime, recognizing that sufficient food output and supply at the national level is not the only requirement of food security. Other factors, e.g., low incomes that lead to inability to access sufficient food, can also lead to food insecurity despite of the ample supply of food at the national level. This indicates that the formulation of strategies, schemes and plans to solve the problem of food insecurity, notwithstanding whether the people are able or unable to support themselves, is of paramount importance.

In Eastern Asia, the governmental policies aimed at enhancing food security include two aspects: pro-poor growth and stabilized food price (C. Peter Timmer, 2004)

Food insecurity does not usually occur in rich countries. The primary reason is not that rich countries have high food reserves or per capita income as large-scale unemployment in these countries can also lead to food insecurity and even famine. The key lies in the fact that rich countries have rather complete social security systems which guarantee the right of everyone to acquire food, whereas it will be unaffordable for poor countries to implement such type of social security systems. Sri Lanka, however, implements a free rice distribution program (Amartya Sen, 1982). Even in low-income countries, intervention by the governments very much promotes food security and alleviates the threat of hunger or even famine. The most successful examples include Eastern Asia and Southern Asia (C. Peter Timmer, 2000).

In coping with food security problems at the household and individual levels, most governments of Eastern Asia adopt the following practices: Improving the educational level of rural women and the poor; Implementing household plans and child care clinics in rural areas; Conducting nutrition education and skill trainings, etc. Most of the documents on food security also concentrate on measures taken by the government at the household and individual levels.

4.2. *Improving agricultural productivity and all other agriculture-related soft and hard infrastructures*

Improvement in agricultural productivity has a noticeable effect on alleviation of poverty and food insecurity. The price of food insecurity is heavy, such as, long-term starvation, malnutrition and even famine. United States Agency for International Development put forward six key steps to improve agricultural productivity. First, the policy framework should be improved. Second, research on and application of agricultural science and technology should be promoted. Third, domestic and international circulation and trade market should be developed. Fourth, property rights should be protected. Fifth, human resources should be improved, such as, improvements in the education and health levels of the population. Lastly, public institutions should be answerable and transparent, as well as law enforcement. These are the basis for the elimination of food insecurity in the short term. High agricultural productivity and poverty reduction policies will help people to break the vicious circle of poverty-starvation-malnutrition. Empirical studies also provide positive proof for benefits of agricultural productivity. In Africa, for example, a 10% increase in the level of agricultural productivity is associated with a 7.2% reduction in poverty. In India, a similar increase in productivity has been estimated to decrease poverty by 4% in the short run and 12% in the long run (Data quoted from the *Agriculture, Food Security, Nutrition and the Millennium Development Goals* by Joachim von Braun).

In order to eliminate factors restricting the ability of the poor to acquire food, the International Food Policy Research Institute (IFPRI) engages in research projects aiming to help poor people on anti-poverty, micronutrient deficiencies, micro-finance, urban food security, gender and development, distribution of family resources, etc. Without proper policies, institutions and rural infrastructure, the agricultural product market will not function well, thus increasing poor people's expenditure on food and reducing their production income. In order to induce the market, trade and supporting organizations to be more inclined to benefiting the poor, IFPRI analyzes the reform of agricultural product market, trade policies, WTO agricultural negotiations, efficiency, actions, diversified income of institutions, activities after harvest, agricultural industry, etc.

IFPRI's IMPACT-WATER model allows it to project the results under various policies and investment situations. Rural road construction, provision of education, supply of clean drinking water, agricultural research and irrigation — these are the five most effective ways to solve the problems of institution, poverty and malnutrition.

4.3. *Promoting economic development and sound governance so as to alleviate poverty in a comprehensive way*

By implementing programs that promote economic development, improve government governance, and solve the problem of poverty, the problem of food insecurity can be comprehensively resolved. (Luther Tweeten, 1999).

5. Perspectives of International Researches on Food Security

The early studies on food security placed their emphasis on promoting food production and increasing the aggregate food supply. The balance between the supply of and the demand for food has been the focus of researches by World Bank and IFPRI, and the role of food reserve in food security was also given special emphasis, such as, estimation of world food reserve by using the Multi-Objective Linear Programming model, comparison of the costs and efficiency of food security under different schemes by using the stochastic model.

As was mentioned above, researches on food availability of the vulnerable groups focused on how to help the food-insecure groups to get through the famine, and the measures to be taken. Researches were carried out to study the relationship between various types of plans and measures and the alleviation of food insecurity. For example, some scholars have constructed models showing that there was a positive correlation between the reduction of civil war frequencies and food security (J. Craig Jenkins and Stephen J. Scanlan, 2001).

Researchers also made investigation on the role of food aid plans in alleviating food insecurity problems of low-income households, such as, the willingness of American low-income residents to participate in the Food

Stamp Program (FSP), a program that aims at helping American low-income households (Helen H. Jensen, 2002; Sonya Kostova Huffman, 2008).

Changes in consumption structures, competition of biofuels for food, global warming, etc., will lead to reduction in food output, affect food security, and exert a more far-reaching influence on developing countries (Joachim von Braun *et al.*, 2007). This is a new direction of researches on food security.

Chapter 5

Analysis on the Causes and Results of the Great Famine of China (1959–1961)

The Great Famine that swept over the whole of China during the 1950s and 1960s, has been the subject of many researches, most of which focused on the estimation of the death toll and the causes of famine. This chapter intends to sort out previous research works and then analyze this famine in terms of food security. This chapter includes three parts. The first part gives a brief account of this famine, including an analysis of the death toll, etc. The second part, which is the focus of this chapter, places emphasis on a comprehensive analysis of the causes of the famine. The last part briefly discusses the measures taken to solve the famine problem at that time.

1. Overview of the Great Famine of China

In terms of famine, the most intuitive research subject is the direct result caused by the famine, i.e., the unnatural death toll. The official data was not released until the early 1980s by the National Bureau of Statistics. Estimations made by scholars were mainly based on the data of the national census in 1953, 1964 and 1982. Therefore, no accurate figure of the death toll exists as yet.

According to the current official statistical data, the national population increased by 12 million in 1959 over the previous year, among which the urban population went up by 16 million while the rural population decreased by four million, which was actually the number of people starved to death during the famine under the rigid dual urban and rural household registration system at that time. The national population decreased by 10 million in 1960 over the previous year, among which the urban population continued to increase by seven million while the rural population continued to decline

Food Security and Farm Land Protection in China

Table 5.1. The Urban and Rural Population During the Great Famine (10,000 persons).

Year	Total population	Urban population	Rural population
1957	64,653	9,949	54,704
1958	65,994	10,721	55,273
1959	67,207	12,371	54,836
1960	66,207	13,073	53,134
1961	65,859	12,707	53,152
1962	67,295	11,659	55,636

China's urban and rural population the Great Famine (10,000 people)

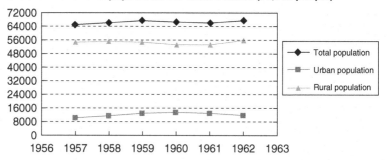

Figure 5.1. China's Urban and Rural Population During the Great Famine (10,000 people). *Source*: Adapted by the Author.

by 17 million, indicating that the famine had not affected the urban areas yet, whereas the rural death toll kept increasing. The national population decreased by 3.5 million in 1961 over the previous year, among which the urban population declined by 3.7 million while the rural population went up by 0.2 million, which, however, does not indicate that the urban population began to bear the consequences of the famine while the rural population started to restore. Please see Table 5.1 and Fig. 5.1.

In fact, the above official data admitted the unnatural death toll of 13.48 million. Regarding this figure, various versions actually exist in both domestic and international academic circles.

According to the statistical data of population, the average population growth rate was 1.91% between 1953 and 1957 and 0.39% between 1958 and

1962. Assuming if, from 1958 onwards, the population grew at the average population growth rate between 1953 and 1957, the national population in 1962 would have been 723.82 million, 50.87 million more than the actual national population in 1962 of 672.95 million. If we calculate the new born population between 1958 to 1962 according to the normal birth rate, and then compare it with the actual newly-born population (to find out the difference), then the unnatural death toll could be calculated.

Table 5.2 briefly cites the estimation of the death toll during the Great Famine by some scholars, ranging from 17 million to over 30 million.

Other representative death tolls are as follows:

According to the official statistics in the *Seventy-Year History of the Communist Party of China* compiled by the Party History Research Office of the Central Committee of the Communist Party of China, the national population of China decreased by over 10 million in 1960.

Years of Tortuous Development, a monograph on the history of the Communist Party, believed that the sum of unnatural death toll and birth decrease was about 40 million.

In *Survival and Development* published by the Science Press in 1989, the Research Group of National Situation Analyses of The Chinese Academy of Sciences believed that a conservative estimation of the death toll due to malnutrition would be about 15 million.

In *A Record of the History of the People's Republic of China* published by the Red Flag Press in 1994, Lu Tingyu believed that the sum of the unnatural death toll and the birth decrease was about 40 million.

During a news release conference in September 2005, Jia Zhibang, Vice Minister of the Ministry of Civil Affairs, said that no corrections would be made to the figure of the death toll between 1959 and 1961.

In contemporary domestic and international academic works, the unnatural death toll of 17 to 30 million was widely adopted whenever the Great Famine of China is referred to. In the entry of "famine" in the relatively conservative Encyclopedia Britannica, the Great Famine of China was referred as the largest famine in the 20th century, "causing as many as 20 million deaths".

Mao Yushi[1] carried out a basic research on the death toll of the famine: the population of a year equals to the population of the previous year plus the number of newly-born population and minus the number of deaths.

Table 5.2. Estimations of the Death Toll During the Great Famine by Some Scholars.

Authors and publication time	Estimations of the death toll during the three years	Basis	Research methods
B. Ashton (1984)	29.47 million		
Ansley J. Coale (1984)	27 million	Three national censuses and the 1/1,000 fertility survey of China	Calculate the numbers of births and deaths for the calendar years between two national censuses, from which find out the linear tendency of death rate between 1958 and 1963; estimate actual death toll, compare it with the linear tendency to find out the difference which exceeds the linear tendency of death rate.
Jiang Zhenghua, Li Nan (1986)	17 million	Three national censuses and the life table of 1981	Establish a dynamic mathematical model, from which find out the annual birth rates and death rates to estimate the annual death tolls; estimate the normal death tolls according to the normal life expectancy, find out the difference between reported death toll with that calculated from normal life expectancy.
Jin Hui (1993)	34.71 million	Official statistical data of population and birth rates	Formula: Births − Deaths + Total Population Decrease = Unnatural Death Toll
Li Chengrui (1997)	21.58 million		Revised the research done by Coale
Cao Shuji (2005)	32.45 million		Using the methods of demography and historical geography to rebuild the unnatural death toll in different regions of China over the years of famine.

The number of deaths includes both the normal deaths and deaths caused by starvation. This could be represented by the formula: the population of the previous year + newly-born population − (natural deaths − abnormal deaths) = the population of this year. The figure of the population before and after the famine could be found out in statistical data. If the newly-born population and the number of natural deaths during the famine can be found out, deaths due to starvation would be worked out. However, both the newly-born population and the natural deaths during the famine are unknown and could only be assumed. If these two figures remained the same as that before the famine, it could be calculated that death toll caused by starvation is 42 million. However, since the newly-born population during the famine certainly would decrease, the actual deaths by starvation could not possibly reach 42 million. Thirty-five million is more credible.

This chapter will not discuss the conflicting figures of the death toll. The figures mentioned above are intended to demonstrate the severity of this famine.

2. Analysis on the Causes of the Great Famine

The focus of this chapter is to explain the causes of this famine. Several causes of this famine have been mentioned, e.g., the natural disaster explanation given by the government which alleged that natural disasters caused substantial fall in grain output; other causes were also mentioned, e.g., it was alleged that people's communes did not function as they should, the over-reporting of grain production and high grain procurement levels, serious waste by the communal dining halls, etc. The following passages will elaborate on the possible causes of the Great Famine in a comprehensive way and attempt to give an unbiased explanation. First, we will introduce the policy background before the Great Famine took place.

2.1. *Background*

2.1.1. *The movement of cooperative transformation of agriculture*

After the completion of the land reform, Mao Zedong envisioned that the next stage of work was to guide the peasants to follow the road of mutual assistance and cooperation, and to transform the dispersed and backward peasants' small private economy into collective socialist agricultural economy.

The earliest form of the cooperative is the mutual assistance team which was composed of four to five rural households whose labor power, farming tools and animals were pooled together for mutual assistance during busy seasons. This type of cooperation did not affect the production decisions of individual rural households based on their own judgments. The higher form was the elementary cooperative which was made up of 20–30 neighboring rural households who completed tasks together according to collective decisions, and were remunerated according to their work. Land, animals and farm implements were still owned privately. The highest form was the advanced cooperative which was usually composed of a village. In advanced cooperatives, complete collectivization was implemented in the form of working points recording; the income of rural households in cooperatives depended on the working points they earned and the value of each working point was determined by the net income of agricultural products of the whole cooperative. In various cooperative movements, peasants were mobilized to engage in rural infrastructure constructions, such as, irrigation, flood prevention, wasteland reclamation, road repairs, etc.

On September 9, 1951, the Central Committee of the Communist Party of China (CCP) held the first meeting on mutual assistance and cooperation of agricultural production and *The CCP Central Committee's Resolution on Mutual Assistance and Cooperation of Agricultural Production (draft)* was promulgated for implementation since December 15, 1951. The draft resolution pointed out that both the enthusiasm for the development of private economy and that for mutual assistance and cooperation existed among peasants after the land reform. In terms of the effects of production and development, the movement of cooperative transformation of agriculture in China was successful, indicating that the agricultural collectivization advocated by Mao Zedong was in line with the productivity development status of China at that time. However, this success did not last long. Stimulated by the impetuosity and rash advance sentiments, all parts of the country started to make big strides forward. By the end of 1954, 480,000 elementary cooperatives had been established throughout the country. By April 1955, the number had increased to 670,000, greatly in excess of the original plan.

Rural collectivization was implemented blindly and rashly, and tense rural situation started to appear and famine was triggered in some regions.

> *The aggregate grain output of Zhejiang Province in 1954 was 14.1 billion Jin and a total sum of 5.1 billion Jin, or 38% of the total, was acquisitioned and procured. The rural grain ration was severely insufficient. According to surveys, each peasant in Zhejaing Province needed an average grain ration of at least 540 Jin each year in normal cases, but the per capita grain ration in 1954 was merely 477 Jin. During the food shortage period between two harvests in 1955, a great number of rural households had run out of grains and serious famine had occurred in several places.*
> — quoted from *Deng Zihui and China's Rural Reform*, p. 436.

In 1955, Mao Zhedong and Deng Zihui, the head of Rural Work Department of CCP Central Committee, expressed different opinions regarding the pace of collectivization. Deng Zihui was criticized as "a woman with bound feet" due to his argument for slowing down the pace of collectivization. As a result, the pace of agricultural collectivization was rapidly accelerated.

By December 1955, there were 1,905,000 agricultural cooperatives throughout China, among which 1,888,000 were elementary cooperatives and merely 17,000 were advanced cooperatives; 58.7% of the total rural households in China had joined in elementary cooperatives and merely 3.9% joined in advanced cooperatives. However, the number of elementary cooperatives and the proportion of rural households joining in elementary cooperatives declined month by month while the number of advanced cooperatives increased with each passing day. By December 1956, there were 756,000 agricultural cooperatives in total after mergers and upgrading, among which 216,000 were elementary cooperatives and 540,000 advanced cooperatives; 8.5% of the total rural households in China joined in elementary cooperatives, whereas the proportion of rural households joining advanced cooperatives soared to 87.8%. The original plan of completing agricultural collectivization in 18 years was actually completed in four years.

2.1.2. *The people's commune movement*

After the completion of the upsurge of rural collectivization in the countryside of China in the latter half of 1955, another "Great Leap Forward" which involved tremendous changes in the relations of production and the economic institution of agricultural cooperatives was started.

The CCP Politburo Conference held in Chengdu in March 1958 passed a resolution called *Opinions on Consolidating Small Agricultural Co-operatives into Big Ones. The Opinions* pointed out that "in order to meet the requirements of agricultural production and the Cultural Revolution, it is necessary to consolidate small agricultural cooperatives into big ones properly and in a planned way wherever conditions permit". Shortly after the conference, the work of consolidating small cooperatives into big ones was carried out in all rural regions of the country, and the "communistic commune", the "collective farm" and the "people's commune" successively appeared in rural areas. In an article titled *Brand-new Society and the Brand-new Man* which was published in Issue 3 of the Red Flag magazine on July 1, 1958, it was clearly proposed that "the cooperative should be transformed into a grass-root organizational unit with both agricultural and industrial cooperation, that is, the people's commune which combines together agriculture and industry". This was the first time when the name of "people's commune" was used. On August 6, 1958, Mao Zedong said that the name of "people's commune" was fine during his inspection of the People's Commune of Qiliying Town, Xinxiang County, Henan Province. On August 9, 1958, during his conversation with leading officials of Shandong Province, Mao Zedong said that "It is good to establish people's communes . . . ." There was a rush for setting up people's communes after this conversation was published in newspapers. During August 1958, the Enlarged Meeting of CCP Politburo was held at Beidaihe and the *CPC Central Committee's Resolution on the Establishment of People's Communes in the Rural Areas* was passed. After the Resolution was transmitted to the lower levels, a vigorous campaign of establishing people's communes was launched. By the end of October 1958, 12,800,000 rural households or over 99% of all rural households in over 740,000 agricultural cooperatives had been reorganized into over 23,600 people's communes. The rural areas of the country had basically been organized into people's communes.

The main feature of the people's commune is "large in size and collective in nature". The so-called "large in size" was shown in the following two aspects. First, the scale of communes was larger than agricultural cooperatives, and the mentality of "bigger is better" was formed in the practical process of consolidation of cooperatives. On an average, a people's commune was composed of 28.5 agricultural cooperatives throughout the country and included over 4,000 rural households in most regions. Some people's communes were even composed of a whole county in 94 counties of 13 provinces, such as, in Henan, etc. Second, the operation and management of the people's commune was larger in scope than that of the agricultural cooperatives. The original elementary and advanced cooperatives engaged in agricultural production, while the people's commune, being a grass-root social administrative organ, engaged in the all-round development of agriculture, forestry, animal husbandry, sideline occupations and fishery, and placed industry, agriculture, commerce, education, and military affairs together under unified leadership and management. The so-called "collective in nature" was demonstrated in three aspects. First, the people's commune was combined with rural grass-root government administration (or the combination of political and social organizations) and the proportion of public ownership was increased. Second, private plots of land, poultry, livestock, and household sideline production of commune members were converted into communal property; thereby the so-called remnants of private ownership of the means of production were eliminated. Third, organizations were put on a military footing, actions were also initiated on a war footing, and life were put on a collective footing; great efforts were made in so-called public welfare undertakings, e.g., communal dining halls; and the distribution system which integrated the wage system with the supply system was implemented.

2.1.3. *The state monopoly for the purchase and marketing of grains*

The severest flood in a century happened in the Yangtze River Basin in 1953, and the reduced output of summer grain crops had become a foregone conclusion. Moreover, wheat crops in the North, especially in the Northeast, suffered serious frost damage in the same year. As a result, the state granary stock was running low. Although the economic situation showed an upward turn during the few years after the founding of the new China, supply fell short of demand and fluctuation in prices was a normal phenomenon under the prevailing conditions where individual rural households were the basic operation units and the scale of production was small. As the main body of economic control, the state mainly made use of agricultural taxes to control the rural economy. This, however, had already called forth a lot of criticism at that time. The central government once explicitly undertook to adhere to the policy of "less requisition and more purchase" in its control over commodity grain and to stabilize the agricultural tax paid in grain at the level of 1952 within a few years. Another important background is that the Chinese government at that time adopted the Stalinist heavy-industry-oriented development strategy; it attempted to transform an agricultural country into an industrial country and to restore and develop itself from the ruins of war, and strived for self-sufficiency in grains. As such, scarce foreign exchange reserves were used to import capital goods. Deng Xiaoping, then the Minister of Finance, also made it clear that there was no extra money which could be used to raise the grain procurement price so as to augment the state grain reserves and to subsidize peasants.

Under such circumstances, even Chen Yun, who once said at the Third Plenary Session of the Seventh CPC Central Committee that "China is an agricultural country, we cannot exploit the agriculture sector for the sake of industrialization", was driven by the idealism of prioritizing industrialization and "to coordinate all the activities of the nation like playing a chess game", and could do nothing but dished out the measure of Unified Grain Procurement and Marketing (UGPM) at the expense of the peasants. The so-called UGPM refers to the unified procurement and marketing of grains by the state, or the establishment of state monopoly of circulation channels by consolidating all the dispersed management and supply sectors. The measure adopted by Chen Yun was "compulsory procurement in the countryside and grain rationing in cities, strict control over private business and adjustment of the internal relationship".

On October 16, 1953, the CCP Politburo approved *The Resolution on the Implementation of Planned Purchase and Planned Supply of Grains* which stated that the UGPM policy is composed of unified grain procurement (abbreviated as "Tonggou"), unified grain supply (abbreviated as "Tongxiao"), strict control over the grain market by the state and the unified administration of grain by the central government; the four policies are mutually interrelated and none of the four is dispensable; all the principles and policies regarding the amount, standard and price of grain procurement and the price of grain supply, must be uniformly formulated or approved by the central government; in compliance with the established principles and policies, local governments should ensure the implementation of UGPM according to local conditions and division of responsibilities.

From then on, the UGPM system was carried out in China for over 30 years and was not abolished until 1985 when the contracted purchasing system was put into effect.

2.2. Causes of the great famine — the aspect of the demand for and supply of grains

This section makes a classification of all the possible contributing factors of the Great Famine and discusses the causes of the Great Famine from the following two aspects, viz., the demand for and supply of grains, including the situation of grain production, the demand for grains, the UGPM policy in the circulation links and the trade situation before the famine, and the aspect of institutions, such as, the People's Communes Movement, the communal dining halls, etc.

2.2.1. Did natural disasters contribute to the famine? To what extent?

In 1960, the Chinese government began to admit the occurrence of the Great Famine between 1959 and 1961 and referred to it as "Three Years of Economic Difficulty". Soon after that, "Three Years of Natural Disasters" became an official explanation of the Great Famine, and it had been used until now. It was stated that the irresistible factor of natural disasters caused severe crop failures and the insufficient supply of grains ultimately led to the famine.

Nevertheless, quite a few scholars expressed their doubts about this explanation. Some came to the conclusion that there was no nationwide large-scale natural disaster during the famine after an analysis of the natural disaster data in the three years (Wang Weiluo, 2001). Some even claimed that the weather during the three years was more favorable than the average year (Jin Hui, 1998). For sure, the explanation of continuous widespread disasters in such a vast country like China seems ill-founded (Lin Yifu, 1990). Liu Shaoqi once pointed out the crisis was "30% natural disaster, 70% human error."

The following passages provide a year-by-year disaster situation of the Great Famine according to the information recorded in the *Report of the Damage Caused by Disaster in China 1949–1995* (hereinafter referred to as *The Report of the Damage Caused by Disaster*) compiled by the National Bureau of Statistics of China and the Ministry of Civil Affairs and published by the China Statistics Press in 1995.

Severe natural disasters with "an unprecedented disaster area in the 1950s" occurred throughout the country in 1959. The affected area reached 44,630,000 hectares (686,500,000 mu) and the damaged area (where grain output dropped by over 30%) was 13,730,000 hectares (*The Report of the Damage Caused by Disaster*, p. 378). The damaged areas accounted for 30.8% of the total affected area which was no higher than that of the average year, but they were concentrated in the main grain-producing areas. The drought area of Henan, Shandong, Sichuan, Anhui, Hubei, Hunan and Heilongjiang accounted for 82.9% of the total damaged area throughout the country. Moreover, various disasters occurred alternately and seriously affected the crop growth.

In terms of the affected area, the natural disasters during 1959 were unprecedented since the founding of the new China. More importantly, a great variety of natural disasters happened alternately in certain regions. Apart from droughts, frosts, floods, wind and hails, other disasters which were rarely seen since the founding of the new China, such as, locusts, Mythimna separatas and mice, also occurred. The population in the damaged areas during the whole year was 80.43 million, 80% higher than the average level between 1949 and 1958, among which Shandong, Hubei and Sichun each had 10 million famine stricken population. A population of 97.7 million suffered spring famine (mainly refers to the people who faced lack of food, including those who fled from the famine-stricken areas to other regions, those who suffered malnutrition diseases, those who went bankrupt to get through the famine, those who sold or gave away their children, and the unnatural deaths), 2.87 times that of the average level between 1949 and 1958.

After the disasters in 1959, extraordinarily serious natural disasters which were rarely seen in the last century occurred in most regions of the mainland except Tibet in 1960. The affected area was 65,460,000 hectares, ranking top in the past 50 years after the founding of the new China, and the damaged area was 24,980,000 hectares (*Comprehensive Statistical Data and Materials on 50 years of New China*, p. 35, compiled by the Department of Comprehensive Statistics of National Bureau of Statistics and published by the China Statistics Press in 1999, hereinafter referred to as *Comprehensive Data on 50 years*). The main disasters included the extremely severe continuous drought in the North and the serious typhoon disaster and floods in the eastern coastal provinces.

The natural disasters in 1960 not only affected a larger area than before but also happened continuously on the basis of disasters in 1959. Therefore, fatal damages were caused. The population in the damaged areas reached 9,230,000 and the population suffering spring famine was as high as 12,980,000 — 3.8 times that of the average level between 1949 and 1958. Another feature was that droughts and floods happened at the same time with striking contrasts. Rainstorms and floods hit certain parts of a province while continuous droughts struck some other parts of the province, which brought about great complexity and difficulty to disaster relief.

In 1961, extraordinarily severe natural disasters happened consecutively for the third year, with an affected area of 61,750,000 hectares. The severity of the disasters was second only to that of the previous year, ranking second in the 50 years after the founding of the new China. The damaged area reached 28,830,000 hectares, ranking top before 1994 (*Comprehensive Data on 50 years*, p. 35), among which one fourth suffered total crop failure (a reduction of over 80% in crop output). In 1991, the population in the damaged area surpassed that of the previous year to reach 163 million and the population suffering spring famine was as high as 218 million, 6.4 times that of the average level between 1949 and 1958 and accounting for over one third of the total national population.

Hence, Li Jinping (2006) believed that the "Three Years of Natural Disasters from 1959 to 1961" was indeed the severest natural disaster in the first 50 years after the founding of the new China; it affected the largest area, made the most serious damages and lasted the longest, and was a direct cause of the three years of economic difficulty. On October 29, 1960, Zhou Enlai remarked at the Enlarged Meeting of CCP Politburo that "Such serious natural disasters have never happened in the past 11 years after the founding of our country, and people at my age have never heard of such disasters ever since we could remember in the 20 century". (Jin Chongji, *Biography of Zhou Enlai*, p. 1558, published by Central Party Literature Press in 1998). Ten most seriously disaster-stricken provinces included Hebei, Liaoning, Jiangsu, Zhejiang, Anhui, Shandong, Henan, Hubei, Guangdong, Sichuan; the most affected three provinces were Shandong, Henan and Anhui (*Major Natural Hazards and Their Reduce Strategies in China (General Introduction)* compiled by the Major Natural Hazards Research Group of the State Commission of Science and Technology, published by the Science Press in 1994, p. 38). In the late 1950s when productivity was low, people's ability to resist natural disasters was limited and the comprehensive national strength was weak, and economic difficulties could not be avoided when extraordinarily severe natural disasters occurred for three successive years.

Figures 5.2–5.4 examine the trends of the affected and damaged areas to see if obvious natural disasters happened in these three years in comparison with other years.

Two indicators are widely used in statistical materials for assessment of agricultural natural disasters: The affected area and the damaged area. The latter refers to the affected area with an output reduction of over 30%

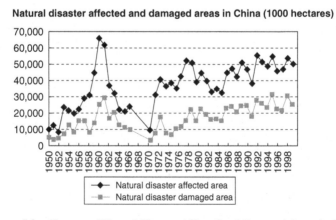

Figure 5.2.   Changes of Natural Disaster Affected and Damaged Areas in China.

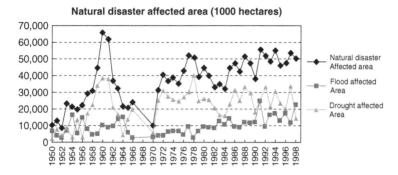

Figure 5.3.   Changes of Flood and Drought Affected Areas in China.

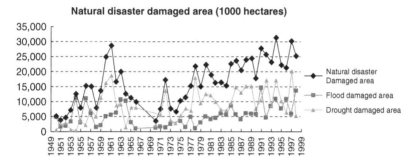

Figure 5.4.   Changes in Flood and Drought Damaged Areas in China.

due to damages of natural disasters. On the one hand, the two indicators should be comprehensively investigated. On the other hand, the damaged area should be specially investigated in assessment of the effects of natural disasters on agricultural production. China was mostly affected by droughts in 1959; in comparison with historical data, the total damaged area in that year was smaller than those of 1956 and 1957, standing in the medium-low level in the historical series of over 50 years. The grain output in 1956 and 1957, however, was both stable, and no nationwide famine occurred. Natural disasters that happened in 1960 and 1961 were also mostly droughts with the damaged areas standing at high levels but not the highest in history. The ratio of the damaged area to the affected area of the three years also stood at the medium-low levels in the historical series, being 31%, 38% and 47% respectively. Death toll was highest during 1960. According to the planting structure of grains in China (summer grains account for about 20%

and autumn grains around 80%), the period when spring is changing into summer (from early March to late May), or the period when grain supplies has run out and the new harvest is not yet in, is the critical period of food security in China when fluctuations in grain prices easily occur and even famine takes place in extreme cases. This judgment is still applicable in China today. In fact, the 10 million people who died in 1960 lost their lives in the first half of the year. Therefore, judging from the time series, even if the natural disasters in 1961 did exert certain effects, they only made the bad situation worse and were definitely not the creator of the unprecedented great famine. Besides, more natural disasters happened in the latter 30 years of the past 50 years after the founding of the new China, and the crucial years include 1978, 1980, 1991, 1994 and 1997. However, no grievous famine ever occurred in these years.

The judgment by Jin Hui (1998) that the weather in the three years was more favorable than the average year was based on an analysis of the *Table of Flood and Drought Grade of Different Regions in China over the Years (1895–1979)*. We believe that this analysis is rather dogmatic and the affected and damaged areas in *The Report of the Damage Caused by Disaster in China 1949–1995* can better illustrate that serious natural disasters did happen in these three years.

In summary, though serious natural disasters did occur from 1959 to 1961, the extent of the disasters was not the most severe in comparison with that of other years; hence the government's attempt to make use of the three years of natural disasters to explain away the Great Famine is ill grounded.

2.2.2. *Grain production*

Because of exaggeration and false report of the output, the grain production situation during the Great Leap Forward could only be estimated. Please see Table 5.3 for the grain production situation before and after the three years of Great Famine.

The grain output fell by 28.3% from 400 billion Jin in 1958 to 287 billion Jin in 1960, and paddy played a decisive role in the sharp decline. The proportion of paddy, wheat and corn in the total grain output was 40%, 11% and 12% respectively. The paddy output decreased by 14%, 13% and 10% respectively in the three years; the wheat output reacted rather slowly with no increase or decrease in the first two years until 1961 when it dropped

Table 5.3. The Output of Grain Crops, 1958–1962.

Year	Total sown area (10000 mu)	Average yield per mu (Kilograms)	Actual total output (100,000,000 Jin)	Year-on-year increase or decrease in total output (%)
1958	191,420	105	4,000	2.5
1959	174,034	98	3,400	−15.0
1960	183,644	78	2,870	−15.6
1961	182,165	81	2,950	2.8
1962	182,431	88	3,200	8.5

Source: *A Compilation of Statistics on China's Rural Economy (1949–1986)*.

dramatically by 35%; the corn output declined sharply by 28% along with paddy in the first year, and the output reduction rate then dropped by 3% in each year of the latter two years. In view of the proportions of the three major grain crops and their output reduction ranges, the production failure of paddy should be the main contributing factor of the occurrence of famine.

Luo Pinghan (2006) provided some feasible explanations for the reduction in grain output. First, the grain sown area decreased substantially[1] because the misreporting of high grain output in 1958 created a false impression that people could sow less but reap more. Second, as mentioned above, the natural disasters in the three years caused serious damages to a large area of the country. Third, insufficient input of labor power, draught animals and fertilizers led to the decline in per mu grain output. For instance, insufficient input of labor power was mainly caused by the mass steel smelting campaign and the large number of recruitment of workers in cities. According to investigations, there were 153 million people working in the forefront of agricultural production in 1957; by the first half of 1960, the labor force was reduced by 33 million to only 120 million. Apart from the

---

[1]Liu Shaoqi inspected both the urban and rural areas of Jiangsu Province in September 1958. During the discussion with the leading officials of the Prefectural Party Committee of Huaiyin City, Liu Shaoqi said that "when I inspected Hebei and Henan, some secretaries of the County Party Committee suggested that less sowing but more harvest is much more economical than wide sowing but meager harvest and the experience of high-yield plots should be spread so as to do farm work better by concentrated utilization of manpower and material resources. In this way, we will be able to grow grain crops on one third of the cultivated land, to plant trees on another one third of cultivated land and to use the remaining one third for recreation in a few years".

excessive recruitment of workers in industrial sectors, the overextension of non-agricultural sectors within people's communes and the excessive transfer of peasant workers by the state are also major reasons for this reduction. The labor force working in the forefront of non-agricultural production was 37 million in 1957 and increased to 89 million in the first half of 1960. Fourth, what is more important, the People's Communes Movement dramatically reduced peasants' incentives for production and it had become a common phenomenon for peasants to show up for the work but without exerting themselves, thus resulting in low agricultural productivity.

2.2.3. *Grain stock*

The severe grain shortage crisis in 1960 directly affected the grain supply in urban areas. A report by the Office of Finance and Trade of the State Council said that grain stock in large cities and industrial regions, such as, Beijing, Tianjin, Shanghai and Liaoning, was very low; Beijing's grain stock was only enough for sale for seven days; Tianjin's stock was only enough for sale for ten days; Shanghai had hardly any rice stock; grain stock in ten cities of Liao Ning was only enough for sale for eight or nine days.

In his letter to Mao Zedong on May 17, 1962, Li Xiannian said that "At present, no more grain can be dug out from the state granaries and the masses too, are running out of grain stock. Grain supply is tight in both grain-surplus regions and grain-shortage regions. Grain stock in several well-known high-yield regions and grain-surplus regions are already emptied and the living conditions of local peasants decline significantly". (*Selected Works of Li Xiannian*, p. 259, published by People's Publishing House in 1989).

There was 34.3 billion Jin of grain reserve throughout the country by the end of June 1959, which was enough to meet the need of urban residents for one year. However, under the grim situation of a substantial grain output reduction of 30 million tons at that time, the government decided to export 4.15 million tons of grain, 121% of the above-mentioned grain reserve, for exchange of gold and dollars. The export of grain exhausted the grain reserve. Soon after that grain supply in such big cities as Beijing, Tianjin and Shanghai was almost out of stock, and hundreds of millions of starving peasants in the countryside had no other options than facing death.

2.2.4. Demand for grain

44.4 billion Jin of grain had been sold from July 1st to the end of December 1958, 9.8 billion Jin higher than the same period of the previous year, among which 5.8 billion Jin was sold in the rural areas and 4 billion Jin sold in urban areas. The main reason for more sale of grains in rural areas was that during the so-called "large formation warfare" organized in rural areas at that time, large numbers of young and middle-aged rural labor force who were amassed to produce iron and steel and to build irrigation works and roads did not bring any grain with themselves and ate public grains wherever they went. The reasons for more sale of grains in urban areas was that the total urban population of the country increased by around 10 million due to the large increase in the number of workers and staff and the natural increase of urban residents, at the same time the grain ration and other grain consumptions in food manufacturing also went up.

— Luo Pinghan, *Several Problems Concerning Grain Production and Sales in the Years 1958 to 1962, Journal of Chinese Communist Party History Studies*, No.1, 2006.

From the above passages, it can be seen that the increased demand for grains was caused both by the increased rations of the peasants and by the substantial increase in urban population caused by rapid industrialization. Due to the rapid acceleration of the population expansion after the founding of New China and the accelerated pace of industrialization over a short period, the urban population expanded dramatically. Under the policy of "Unified Procurement and Marketing of Grains", the problem of grain shortage was manifested in the relation between the state and the peasants. From the perspective of the peasants, the state procured large amounts of grains even when they were short of rations. Under the condition of extremely low level of per capita grain supply, the forcible launching of industrialization enlarged the demand for grains by the cities so greatly that it contradicted with the demand for rations by the peasants. When grain output was significantly reduced, grain shortage in rural areas was definitely the first consequence of grain procurement by the government. Some scholars believe that the mainstream description of this period in the history of New China is to attribute the economic difficulties to "the reduced grain output caused by "the wrong command of people's communes" and "the high procurement of grains due to the 'wind of exaggeration'" at that time. However, this description was supported by very few actual cases. In fact, the amount of grain corresponding to the urban population would be procured regardless of whether the wind of exaggeration existed or not.

Another important feature of the Great Leap Forward in 1958 was the influx of large numbers of people to the cities, which caused population explosion in cities. By 1960, the urban population in China reached 130.73 million, accounting for 19.8% of the national population. This proportion was not exceeded until 1981. The urban population went up by 38.88 million from 1956 to 1960, among which 10.34 million or 26.6% belonged to natural population growth and the remaining 28.54 million or 73.4% was caused

Table 5.4. Changes in Urban Population in China, 1956–1963.

Year	Urban population (10,000 persons)	Proportion of urban population in total Population (%)	Sources of urban population increase (10,000 persons)			
			Organic changes	Migration increase	Natural Growth	Total
1956	9185	14.6				
1957	9949	15.4	20	386	358	764
1958	10721	16.3	90	421	261	772
1959	12371	18.4	−60	1, 481	229	1, 650
1960	13073	19.8	200	316	186	702
1961	12707	19.3	90	−614	158	−366
1962	11659	17.3	−140	−1, 225	317	−1, 048
1963	11646	16.8	−170	−278	435	−13

Note: (1) Urban population in this table refers to population living in the administrative areas of cities and towns (excluding the governed counties).

(2) Organic changes were calculated on the assumption that each city has a population of 0.1 million.

(3) Natural population growth was calculated by using the natural growth rate of city population.

Source: *National Bureau of Statistics of China, China Statistical Yearbook 1991*. Beijing: China Statistics Press, 1991, p. 79.

by other reasons. Besides, the increased number of cities was also one of the major reasons for the increased urban population (see Table 5.4).

2.2.5. *Grain circulation link: Grain procurement*

Due to the over-reporting of grain output and a bumper harvest in 1958, high grain procurement was implemented in three consecutive years from 1958 to 1960.

Although the actual grain output in 1959 was merely 340 billion Jin, grain procurement in this year was as high as 134.81 billion Jin (67.405 million tons) and the procurement rate was as high as 39.7%. 102.1 billion Jin of grains was procured throughout the country in 1960, lower than that of the previous year. However, due to the decrease in total grain output, the grain procurement rate in this year was still as high as 35.6%. The levy amount of agricultural tax from peasants in 1958 was originally set at 41.2 billion Jin and then raised to 42.8 billion Jin. 44.57 billion Jin of grains was actually procured in that year, an increase of 11.4% over that of 1957. Agricultural tax accounted for merely a small part of the grains turned over by the peasants, whereas the unified procurement of grains accounted for a much larger proportion. When the amount of grains resold to peasants was deducted from the amount of grain procurement, the remainder was the net grain procurement. The net grain procurement rate during the years 1958–1960 (see Table 5.5) surpassed 20% mainly because of the increased demand for grains by a substantially expanded urban population at that time and also because of wastage of grains. For example, the urban population of China in 1958 increased by 7.8% over the previous year while the net grain procurement amount rose by as high as 23.2% (Li Ruojian, 2001).

The excessive procurement of grains brought great difficulty to peasants' life. In 1960, each individual rural person originally occupied 212 kilograms of grains on the average and the per capita rural grain occupation in many places was even less than 200 kilograms. After a deduction of seeds, the per capita ration was even less. The per capita rural occupation of grains after grain procurement in Guizhou Province was 152 kilograms in 1959 and 111 kilograms in 1960 when 1.1 million tons of grains or 68.75% of the procurement amount had to be resold to peasants. It is clear that excessive grain procurement and large amounts of resold grains also led to great waste.

Table 5.5.   The Amount of Grain Procurement, 1958–1962.

Year	Grain output (10,000 tons)	Grain procurement (10,000 tons)	Net grain procurement (10,000 tons)	The ratio of procurement to output (%)
1958	20,000	5876.0	4,172.5	29.4
1959	17,000	6740.5	4,756.5	39.7
1960	14,350	5105.0	3,089.5	35.6
1961	14,750	4047.0	2,580.5	27.4
1962	16,000	3814.5	2,572.0	23.8

Note: The procurement amount was calculated on yearly basis, that is, from April of one year to March of the next year. Net grain procurement was calculated by deducting the resold grains from the total grain procurement.

2.2.6. *Grain trade*

In the 1950s, international grain trade was carried out either to import grains to cover up shortage or to export grains for getting foreign exchange or importing machinery equipments, i.e., for the primitive accumulation of industrial capital. The total grain export of China in 1953 was 3.2 billion Jin, among which two billion Jin of soybean was exported to the Soviet Union mainly for exchange of machinery, 0.54 billion Jin to Ceylon (presently Sri Lanka) for rubber and the remainder to other countries.

The main varieties of grains exported to other countries in the 1950s were rice from Southern China and soybean from Northeastern China. China started to export rice in 1950 and its rice export reached 1.67 million tons in 1956, accounting for 63% of the total grain export in that year, 1.77 million tons in 1959, and then gradually declined. China began to export soybean in 1950 and its soybean export reached 0.91 million tons that year, accounting for 74.5% of the total grain export for that year. Soybean export exceeded one million tons in 1951 and then fell slightly; soybean export from 1955 to 1960 was over one million tons, accounting for 40% to 50% of the total grain export.

Not much wheat was imported by China from the international market in the 1950s, generally less than 0.1 million tons each year, except for 1958, when 0.1483 million tons was imported, accounting for two thirds of the total grain import for that year. Due to the three years of famine, China became a net importer of grains since 1961 with wheat as the main imported

Table 5.6. Grain Import and Export, 1958–1962.

Year	Total grain export (10,000 tons)	Total grain import (10,000 tons)	Net grain import (10,000 tons)	Net wheat import (10,000 tons)
1958	288.34	22.35	−265.99	14.83
1959	415.75	0.20	−415.55	
1960	272.04	6.63	−265.41	3.87
1961	135.50	580.97	445.47	388.17
1962	103.09	492.30	389.21	353.56

grain variety. Wheat import by China continued to increase since the 1960s and rapidly reached 3.88 million tons in 1961. There was no international trade of corn in China prior to 1955. The net import of corn by China started to rise from 1956, it was maintained at below 0.1 million tons prior to 1961, but went up to 0.49 billion tons in 1962.

China's grain export in 1959 reached 4.1575 million tons, the peak of the 1950s, while the grain import during this year was the least, only 2,000 tons. The famine from 1959 to 1961 was also a turning point in China's international trade of grains, when it became a net importer, in contrast to 1950s, when it was a net exporter. The trade surplus in grains since the founding of New China was completely changed since then (see Table 5.6).

2.2.7. Institutional factors — urban-biased food supply system

Lin Yifu and Yang Tao (2000) pointed out that "China has an effective, urban-biased ration system in which city residents were given legally protected rights to acquire a certain amount of food". The government imposed compulsory grain procurement quotas on the peasants according to the needs of the urban residents. Therefore, whatever the grain output was in each year, this system enabled urban residents to enjoy priorities in the entitlement to food, whereas peasants had the right only to the remainder grain supply. A citizen's entitlement to food was decided by his or her identity. Therefore, the entitlement to food was represented by the urban-biased system of procurement of grains.

The above Table 5.7 shows that the difference between the urban and rural per capita grain occupation is as high as 100 kilograms, indicating that

Table 5.7. Grain Procurement Rate, 1958–1962.

Year	Output (million tons)	Net procurement rate (%)	Rural per capita grain occupation (kilograms)	Urban per capita grain occupation (kilograms)
1958	200	13.6	311	228
1959	170	28	223	380
1960	143.5	28	191	308
1961	147.5	17.5	225	274
1962	160	16.1	234	296

Source: Xin Yi. *Research on the Distribution System of People's Communes in Rural Areas*. Chinese Communist Party History Publishing House, 2006, p. 50.

food supply in the urban areas was much better than that in the rural areas during the three years of famine.

2.2.8. *Iron and steel production campaign*

The general line of socialist construction "Go all out, aim high and achieve greater, faster, better and more economical results in building socialism" was officially adopted at the Second Plenary Session of the Eighth Central Committee of the CPC held in May 1958. Soon after that, the communist party launched the Great Leap Forward Campaign and requested the output of main industrial and agricultural products to be doubled, to be increased by several times or even scores of times. For example, in the Politburo meetings in August 1958, it was decided that the steel output of 1958 was expected to double that of 1957, to increase from 3.35 million tons to 10.7 million tons, and that the steel output of 1959 was expected to double that of 1958, to increase from 10.7 million tons to 30 million tons.

In November 1957, Mao Zedong proclaimed that China might overtake and surpass Britain in the output of steel and other major industrial products in 15 years. Encouraged by the slogan of "steel as the key link to the full leap forward", the production quotas of iron and steel was raised up higher and higher. At the Beidaihe Politburo meetings, it was officially decided and declared that the nation must turn out 10.7 million tons of steel in 1958 to double the 1957 figure and that the whole party and people were urged to strive for this goal by plunging in the unprecedented iron and steel production campaign. The major practices were: (1) All departments and regions were to prioritize production and construction of iron and steel in the first place so as to make way for the national iron and steel production campaign; (2) The first secretaries of the party committee at all levels were to take command in unfolding mass movements to produce steel using indigenous methods; (3) Additional investment was made continually to expand the production capacity of existing enterprises, leading to the rapidly expanded scale and overextension of capital construction; (4) Commercial banks rendered full support to the Great Leap Forward in industry, resulting in "robbing Peter to pay Paul" and normal circulation of capital was disrupted. After foolhardy efforts, it was declared on 19 December 1958 that the task of doubling the steel output had been fulfilled 12 days earlier than the plan with the steel output reaching 11.08 billion tons and the output of pig iron at 13.69 million tons. In reality, only 8 million tons of steel was of acceptable standards, and the three million excess tons of steel and 4.16 million tons of iron produced using indigenous methods could not be used at all. It was estimated that 20 billion RMB was lost in iron and steel production throughout the country during this period.

A large number of rural labor forces were engaged in production of iron and steel, whereas insufficient labor forces were retained in rural areas for agricultural production. As a result, some ripe crops were left to rot in the fields causing a significant reduction in grain output. In Henan Province, where the inclination for exaggeration was most rampant, only 50% of cotton and 70% of sweet potatoes and peanuts had been harvested, and only 60% to 70% of the harvested crops were threshed by 6 November 1958. In Sichuan Province, it was estimated that grains left scattered in the fields was more than 2.2 million tons, accounting for 10% of the total grain output. In Shaanxi Province, more than 1.8 million rural labor forces were enlisted to engage in production of iron and steel in 1958 and only the old, weak, and women and children were left to do the autumn harvest. This approach resulted in a reduction in grain output. In Dexing County of Jiangxi Province, only 10% of the 120,000 mu of the late rice was gathered and 27% of the winter planting plan of 179,000 mu was completed in 1958 because more than 10,000 of rural labor forces were enlisted to engage in production of iron and steel; by 1960, 47,000 people or 29.7% of the county population (158,000) had left the county.

2.2.9. *Wind of exaggeration — exaggeration of grain output*

Although the farmers did have a good harvest of grain in 1958, the grain output was excessively estimated due to the Wind of Exaggeration. In August 1958, the Enlarged Meeting of CCP Politburo at Beidaihe estimated and officially declared that the grain output of 1958 would reach 300 to 350 billion kilograms, 60% to 90% higher than that of 1957. At the end of 1958, relevant departments "further exaggerated the grain output to 425 billion kilograms" based on the reports by different regions (The report by Tan Zhen Lin and Liao Luji on *Major Situation, Problems and Suggestions Regarding Agricultural Production and Rural People's Communes* approved and transmitted by the CPC Central Committee on December 7, 1958).

The Communiqué of the Eighth Plenary Session of the Eighth Central Committee held on August 16, 1959 pointed out that "the previously released statistics of agricultural output in 1958 was on the high side" and believed that after verification the actual grain output should be 250 billion Jin, which was, however, still far higher than the actual output. In reality, a further verification proved that the national grain output of 1958 was merely 200 billion kilograms, much lower than the released figure. The seriously erroneous estimation gave people the false impression that there were so much grains that there would be excess after feeding the population. Nevertheless, the National Economic Plan For 1959 proposed by the State Council was approved at the First Session of the Second National People's Congress held in April 1959. The plan proposed that "the mission for

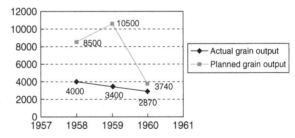

Figure 5.5. The Difference Between the Actual National Grain Output and the Planned National Grain Output.

Source: Compiled Using Historical Materials.

Table 5.8. Exaggeration of Grain Output in Different Regions.

Region	Exaggerated output (10,000 tons)	Real output (10,000 tons)	Exaggeration rate (%)
Gansu	1,000	339	195.0
Henan	3,510	1,265	177.5
Hebei	2,250	837.6	168.6
Anhui	2,250	885	154.2
Hubei	2,250	987	128.0
Jiangsu	2,000–2,500	1,110	80.1–125.2
Zhejiang	1,600	789	102.8
Sichuan	4,500	2,246	100.4
Fujian	885	445.5	98.7
Guangxi	2,290	1,170	95.7
Qinghai	110	58.6	87.7
Hunan	4,500	2,455	83.3
Guizhou	900	525	71.4
Shanxi	750	462	62.3
Inner Mongolia	590	423	39.5

Source: *Local chronicles, Contemporary China Series (published by Contemporary China Publishing House), etc.* Due to the large number of reference books, we did not list the titles of all of these books.

agricultural development in 1959 is to try our best to bring the grain output to 1,050 million Jin". The Grain Plan for 1960 approved by the Second Session of the Second National People's Congress aimed to increase the grain output of 1960 by 10% over the actual output of 1959, that is, 374 billion Jin. Figure 5.5 and Table 5.8 shows the difference between the actual grain output and the planned output as well as the exaggeration of grain output in different regions.

The People's Daily published on 23 July 1958 an editorial titled "What does the bumper harvest of this summer indicate?". Steeped in the typical style of language of the Great Leap Forward period, it proclaimed that "China's wheat output has surpassed that of the United States to rank second in the world. The growth rate of wheat output in our country was unprecedented in history of modern or ancient times, in China or elsewhere, and leaves all the capitalist countries treading behind in the dust". "The U.S. imperialists said that the large population of our country is 'an unbearable pressure'.

We will tell them with more facts that, a larger population produces more grains. So long as we need it, we can produce as much grains as wish". Later, it also promoted the slogan, "How much courage people have, how much the land will yield", "The yield of the land is determined by the courage of the people".

The People's Daily reported on 12 July 1958 that: according to the report dispatched from Zhengzhou on 11th July by Xinhua News Agency, the total output of the experimental plot of two mu in the No. 4 Production Brigade, Hepin Agricultural Commune, Chengguan Town, Xiping County, Henan Province reached 14,640 Jin and the per mu yield is 7,320 Jin. This is the twenty-ninth "satellite" with per mu wheat yield of more than 3000 Jin in Henan Province's wheat harvest this year.... The two mu of wheat grew especially well with large and long ears, plump seeds and dense but evenly-distributed plants, namely, 1,486,200 plants per mu on average. The crops are so thick that even a mouse cannot squeeze through them. The largest ear of wheat had 130 seeds; the average ones had seventy to eighty seeds; even the smallest one had around fifty seeds. The two mu of wheat was reaped on June 18th, weighed and garnered in on July 1st.

A "big satellite" of per mu paddy rice yield of 130,000 Jin was set by Huanjiang County in northwest Guangxi, where a number of minor nationalities lived together, on 9 September 1958. This was the biggest false satellite regarding paddy yield in the country at that time.

It should be noted that even in the science circle, where a rigorous and pragmatic attitude should have been adopted, exaggeration phenomenon also existed, and some scientists even theorized the feasibility of high grain yields. Qian Xuesen wrote an article titled *How Much Grain Will a Mu of Field Produce?* which was published in China Youth Daily on June 16, 1958. In view of the per mu wheat yield of two or three thousand Jin in Weixing Agricultural Commune, Suiping County, Henan Province, Qian Xusen wrote, "Has the grain output of the land hit the ceiling? Scientific calculations tell us, far from it!" Then he predicted that the yield per mu of paddy and wheat would be "more than twenty times that of two thousand

Jin!" Besides, when Mao Zedong asked "what should we do when there is too much grain", the Chinese Academy of Sciences assigned urgent tasks on finding ways of transforming grains into industrial raw material.

The wind of exaggeration in the scientific circles gave scientific grounds to the false grain outputs. The support given by the scientific circles for satellites of grain output added fuel to the flames of exaggeration in grain outputs.

2.2.10. *"Communist wind" — communal dining hall*

Due to the "communist wind" in the movement of people's communes, peasants had to turn over all their means of subsistence and the small amount of means of production to the collective, and their individual reserves were all swept away. The gap between the rich and the poor was bridged and some people were deprived of their possessions by some other people. The practice of equalitarianism had led to common poverty.

According to the statistics provided by the National Bureau of Statistics, in January 1960, 399,000 communal dining halls had been set up successively in the rural areas of the country and 400 million people ate at these communal dining halls, accounting for 72.6% of the total population attending people's communes.

In his monograph on the Great Leap Forward, Yang Dali believed that food in the communal dining halls as well as land and farm implements in the communes were all seen as public properties; since meals in the communal dining halls were supplied for free and without any quantity restriction, peasants ate as much as they could; the benefit of eating more food was owned by individuals while the cost was shared by other members of communes; that was how "the Tragedy of the Communes" happened.

The communal dining hall, which was perceived as an element of communism, actually caused tremendous waste of food and great ideological confusion.

Eating free meals" and "eating as much as possible" also caused tremendous waste of food in communal dining halls. Some commune members ate as much as they could in each meal, worrying that they would be at a disadvantage if they ate little, and even secretly brought food which they could not eat any more back home to feed chickens and ducks. It is sheer nonsense to say that "eating as much as possible" can save food. In many communal dining halls, two or three Jin of grain was eaten up by each person on average. In the communal dining hall of Duanzhuang Production Team, Diaotun Production Brigade, Malinggan Commune, Heze County, Shandong Province, 340 people ate up 10,000 excess Jin of grain in one month after they were allowed to "eat as much as they could," equaling to three Jin per person per day. This is not an unusual phenomenon. Five meals were served in some places, and meal "satellites" were even sent in some places. In this way, after the policy of "eating as much as possible" was implemented for two or three months, most communal dining halls had began to eat the grains stored up for the next year.

— quoted from the Big Rice Bowl: the Story of Communal Dining Halls. Luo Pinghan, Guangxi People's Publishing House.

2.2.11. *The Soviet Union pressed for payment of debts?*

The Soviet Union pressing for payment of debts was also one of the official explanations for the Great Famine. Nevertheless, according to the historical record, this explanation does not hold water.

The Soviet Union turned against China in June 1960 when the Soviet Union unilaterally terminated its agreement to provide China with economic aid and withdrew its experts from China. In return, China offered to pay off its debts to the Soviet Union ahead of schedule by using part of its trade surplus with the Soviet Union. However, by 1960 the great famine had already occurred.

According to Zhou Enlai's *Report on the Work of the Government at the First Session of the Third National People's Congress* in 1964 and Li Xiannian's *Report on the National Budget and Final Accounts at the Fourth Session of the Second National People's Congress*, China's loans from Soviet Union plus interests payable totaled 1.406 billion new rubles, or 5.29 billion RMB, most of which were incurred by military supplies used in the Korean War. According to the agreement, these debts were payable by 1965.

On February 27, 1961, Khrushchev offered to provide China with 1 million tons of grains and 0.5 million tons of sugar in the form of low interest loans. China only accepted the latter offer. *The Sino-Soviet Trade Meeting Communiqué* on April 8, 1961 stated that, "The Soviet Union expresses its full understanding, and proposed to the Chinese government to pay the above-mentioned debts interest free in five yearly installments. The Soviet Union also offers to loan China, interest free, 0.5 million tons of sugar by the end of August this year; China will return this loan of sugar between 1964 and 1967. China accepted this offer with gratitude". (The report of Xinhua News Agency on April 8, 1961, first published in People's Daily on April 9, 1961).

Although Sino-Soviet differences had aggravated previously and the Soviet Union had terminated several military aids to China, however, the economic and technical aids to China by the Soviet Union had never stopped. China and the Soviet Union even signed an agreement on scientific and technical cooperation in October 1959. In his report in March 1960, Li Chufu claimed that the Soviet Union continued to provide China with immense aids in 1959, which played an important role in sustaining the Great Leap Forward in national economy. It is clear that attributing the Sino-Soviet split as one of the causes of the Great Famine makes no sense.

2.2.12. *The Lushan Conference*

The Lushan Conference was an important event in history. The Enlarged Meeting of CCP Politburo was held in Lushan from July 2 to August 1, 1959. The original objective of the conference was to rectify the mistakes in the Great Leap Forward. However, after Peng Dehuai had written a private letter to Mao, the conference process was twisted to expose and criticize the so-called "rightist tendencies" and "Anti-Party Clique" of Peng Dehuai, Huang Kecheng, Zhang Wentian, Zhou Xiaozhou, etc.

After the Lushan Conference, the anti-rightist atmosphere pervaded both inside and outside the Party. Impetuosity and rash advance started to prevail again, and quotas in different regions were not brought down at all. Instead, a new round of "Leap Forward" started.

Table 5.9 summarizes the above-mentioned factors. In general, natural disasters and some human factors resulted in the decline in grain output, which was, however, insufficient to explain the nationwide great famine.

Table 5.9. Summary of the Causes of the Great Famine.

Causes of the Great Famine	Effects on the formation of famine
The Aspect of the demand for and supply of grain	
I Did natural disasters contribute to the famine? To what extent?	
II Grain Production	+
III Grain Stock	
IV Demand for Grain	+
Grain circulation link: grain procurement	+
Grain trade	+
Institutional factors	
I Grain procurement	+
II Urban-biased food supply system	+
III Iron and steel production campaign	+
IV the wind of exaggeration	+
V the communal dining hall	+
VI the Soviet Union	/
VII Lushan Conference	+

Note: "+" means a significantly positive effect on famine, "-" means a significantly negative effect on famine, "/" means no apparent effect on famine.

Some institutional factors, such as, the system of urban-biased Unified Purchase and Marketing of Grain, grain wastage caused by the communal dining halls, the Iron and Steel Production Campaign, the grain procurement quotas higher than that of previous years due to the Wind of Exaggeration and the decision of exporting instead of importing grains when grain production was seriously insufficient, played important roles in causing the Great Famine. These factors will be more convincing after stepwise exclusion and verification is empirically conducted. However, this is not the focus of the discussion in this chapter.

3. Solutions to the Famine

3.1. *Close attention paid to the ordnance transportation of grains*

In view of the substantially reduced grain stock and tight grain supply in Beijing, Tianjin, Shanghai, Liaoning, etc., after the beginning of the spring of 1960, the CPC Central Committee decided to set up a headquarter to deal with the ordnance transportation of grains, oil and cotton, led by Li Xiannian, Vice Premier of the State Council and Minister of Finance, and participated by principals of the State Economic Commission, the Ministry of Railways, the Ministry of Transport and the State Administration of Grain. On February 18, 1960, Li Xiannian made suggestions to the CPC Central Committee and Mao Zedong on taking the following emergency measures to carry on an urgent transportation of grains, oil and cotton: (1) Instructing that departments of railways, and departments of transportation of relevant provinces and autonomous regions, that they must guarantee to fulfill the ordnance transportation plan of grains set up by the CPC Central Committee on time; (2) Giving priority to the ordnance transportation of grains during the period of urgent ordnance transportation of grains, when even the grains and oil for export of international trade should temporarily give way to the ordnance transportation of grains; (3) Allocating the planned number of trucks to provinces and autonomous regions in advance; (4) Organizing necessary sideline transportation teams to join in the urgent ordnance transportation apart from the full-time transportation teams of people's communes. On February 21, the CPC Central Committee approved and transmitted Li Xiannian's report and requested all regions to carry out an immediate study on and implementation of this report; it was also proposed that "all provinces which are assigned to turn out grains should urgently be mobilized, launch a rush campaign of ordnance transportation of grains, and ensure to accomplish the present ordnance transportation mission of grains."

On April 19, 1960, the CPC Central Committee approved and transmitted the *Report on Several Measures of Rush Ordnance Transportation of Grains by the Office of Finance and Trade of the State Council*, requesting Party committees, departments related to grains and transportation of all provinces, municipalities and autonomous regions that they must try their best to take all necessary measures to ensure the accomplishment of second quarter ordnance transportation assignments of 5.6 billion Jin of grains within 50 days to two months before the new grain supply was available in the market, and that transportation of all the other supplies should temporarily give way to the ordnance transportation of grains. Practically in every month from then onwards, the CPC Central Committee would issue emergency instructions on the ordnance transportation of grains, requiring the first secretaries of the Party committees of all provinces, municipalities and autonomous regions to inspect the ordnance transportation of grain by themselves and to ensure the accomplishment of the ordnance transportation assignments of grains in each month. Grain stock at that time was very low, and the only thing that could be done was to transport the early-maturing grains in the South to the North in summer and autumn and to transport the late-maturing grains in the North to the South in winter and spring so as to meet the most urgent needs.

3.2. *Lowering the ration standards in both rural and urban areas*

On September 7, 1960, the CPC Central Committee issued Instructions on *Lowering the Ration Standards in both Rural and Urban Areas* which stipulated that the ration standard was 360 Jin of raw grains per person per year in areas South of the Huaihe River, while that in the disaster-stricken areas would be even lower. In places with good harvests, on conditions that the original assignments and additional assignments of grains to support the disaster areas were fulfilled, and if there were still excess grains remained, then the ration standard could be raised to 380 but no higher than 400 Jin of raw grains per person per year. The rural ration standard in areas North of the Huaihe River should be lowered to below 300 Jin per person per year. In reality, it was impossible for rations in many rural regions to reach this standard at that time. In 1960, since the reduction in grain output had become a foregone conclusion and rations of peasants had been lowered to the extreme that they could hardly sustain their lives, it was impossible to ensure the food supply to urban residents through increased grain procurement. The only way was to lower the quotas of retained grains for the peasants, at the same time lowered the food supply standard to urban residents correspondingly. Therefore, the Instructions by the CPC Central Committee on *Lowering the Ration Standards in both Rural and Urban Areas* clearly stated, "The ration standard of all urban population must be lowered by around two Jin (commercial grain) per person per month except for workers and staffs who work in high temperature environments, high above the ground or down in the mines and other heavy physical laborers."

Through the measures of verifying the urban population, lowering the ration supply standard and deduction of unreasonable food supply, a reduction of 5.3 million Jin of grain supply was realized in Beijing per month, and the per person ration was lowered by three Jin per month.

3.3. Policy rectification: Seven thousand cadres conference

The Seven Thousand Cadres Conference, started on January 11, 1962, was attended by around 7,000 cadres, including the cadres of five levels (central, central bureau, provincial, prefectural and county levels) and the directors and secretaries of the Party committee of major factories and mines. The notice of this conference made it clear that the aim of this conference was to examine the errors and mistakes since 1958 rather than to summarize the experiences and lessons in resisting "natural disasters." Neither the written report of Liu Shaoqi which was discussed and approved by the working conference of the CPC Central Committee nor the long speech by Mao Zedong touched on the issue of "severe natural disasters" but focused on "serious errors in the work". In his impromptu speech, Liu remarked explicitly that it was "30% natural disaster, 70% human error"; Mao Zedong vaguely admitted the "blindness" and talked of taking responsibilities instead of blaming the adverse weather conditions.

The conference decided to abolish the planed grain procurement of 15 billion Jin and to slow down the growth rate of industrial development so as squeeze out some foreign exchange originally planned for industrial purposes to import more grains so as to alleviate local procurement pressure. Besides, the conference also decided to decrease the urban population by a great deal so as to reduce urban food supply and alleviate the grain procurement pressure in rural areas.

3.4. Reducing urban population

Large-scale reduction in the number of workers and staff and urban population was carried out throughout the country in 1961. In April 1961, the CPC Central Committee approved and transmitted *The Report on Adjusting Rural Labor Force and Reducing and Transferring Workers and Staff to the Countryside* by the Committee of Five of CPC on Reducing

Cadres and Allocating Labor Force, and pointed out that "Making adequate arrangements for labor force in both urban and rural areas and reduction of workers and staff is an issue of the utmost importance in current national construction work", "Adequate arrangements for labor force from all walks of life and the reduction of workers are important issues of national construction under the present situation". On June 28, the CPC Central Committee issued *The Circular on Several Issues Concerning the Work of Reducing Workers and Staff* which stipulated targeted organizations to be streamlined, treatment of the streamlined people, resettlement of streamlined people after they return to their native place, matters requiring attention in the reduction of workers and staff, etc. Hereafter, the CPC Central Committee and the State Council gave many instructions on issues concerning the reduction of workers and staff and the urban population. From January 1961 to June 1963, 18.87 million workers and staff were streamlined throughout the country and the rural population was reduced by 26 million.

3.5. *Food import*

Since the domestic food stock had no more potentiality to be exploited, Chen Yun put forward the suggestion of importing food from foreign countries at the end of 1960. Nevertheless, the action of making such a suggestion in itself involved great political risks, which could be seen from the following passage.

> *Since liberation, China had exported grain every year. Moreover, during the "Great Leap Forward", we had declared that we had already solved the food problem long ago, and even said that China could support the population of the whole world. Now we make a 180-degree turn and suddenly want to import food. You can imagine what influence would be exerted both inside and outside the Party and the country. Therefore, eating imported food was taboo at that time.*

> — *quoted from Seminar on Economic Thoughts of Chen Yun*, p. 298.

Despite that, Chen further clarified his considerations for proposing import of food at the working conference of the CPC Central Committee on January 19, 1961. The CPC Central Committee decided to start importing food from 1961.

After the policy of importing food was established, Chen Yun, Zhou Enlai and the principals of the relevant department related to grains and foreign trade together discussed the amount of food import in several meetings, and gradually increased the amount to be imported. Rice import was firstly set at 1.5 million tons at the end of 1960. Then, Chen Yun strived to increase the figure to 2.5 million tons based on the degree of disasters and the tight food supply. In 1961, after an analysis of the quotations of the food market, Zhou Enlai suggested that four million tons of food could be imported, which was implemented by Chen Yun immediately. At the working conference of the CPC Central Committee on March 1961, it was decided that five million tons of food would be imported each year. Food import in the following five years was generally stabilized at this level.

In those years, countries from which China imported food were also selected. Although the U.S. was a big supplier in world food market, it seemed everybody deliberately avoided mentioning it as an option. Food was imported from Canada, Australia and France rather than from the U.S. At the working conference of the CPC Central Committee held at Lushan in August 1961, Chen Yun asked Mao Zedong whether we could purchase food of the U.S. through entrepot trade via France with which China had a close relation. This suggestion was confirmed by Mao. Soon after that, food of the U.S. arrived at China steadily through entrepot trade via France.

China imported around 5 million tons of food annually from 1961 to 1965. Although the food import did not account for a large proportion of the aggregate food amount of the country at that time, as a bulk food supply controlled by the state, it played an important role in the scheduling of food. It stabilized the market and, in particular, helped to provide food in times of emergency. For example, 2.15 million tons of food (4.3 billion Jin) was transported to China in a short time before June 30, 1961, which avoided the danger of grains being sold out in Beijing, Tianjin, Shanghai, Liaoning and the worst-hit areas. Besides, the food import also supplemented part of the national food stock and supported the adjustment and restoration of national economy at that time to a certain extent.

3.6. Rectification on the people's commune system

The three years of famine also exposed the defects of the people's commune system. The government started to intentionally adjust the people's commune system from 1960.

In November 1960, the CPC Central Committee clearly stipulated in *The Urgent Letter of Instruction Concerning the Current Policies for Rural People's Communes'* (abbreviated as *The Twelve Articles*) that people's communes should institute the system of "a three-level ownership with the contingents as the basic units"; the production contingents should be set as the basic owners and their subordinate production groups (with a scale of the elementary cooperative and renamed as production teams after March 1961) also had a small part of ownership; the mistake of equalitarianism and indiscriminate transfer of resources should be firmly opposed and completely rectified. Commune members should be allowed to till small private plots and engage in small-scale household sideline production. Rural fairs should be resumed in a planned way and under proper leadership. The excessive centralization of authority was partially delegated to individuals and peasants began to have small-scale management right and decision-making power.

At the working conference of the CPC Central Committee in May 1961, drastic revision was made to the contents regarding the communal dinning system and the supply system in the draft version of the *Regulations on the Work of the Rural People's Communes* (known as *The Sixty Articles*), which was then transformed into *The Revised Draft of the Regulations on the Work of the Rural People's Communes*. It was stipulated in the revised draft that the communal dining system was to be abolished and the ration system too was basically abolished.

In February 1962, the CPC Central Committee issued *The Instructions on the Alteration of the Accounting Units of the People's Communes* which pointed out that production teams should be used as the basic accounting units of people's communes and that a three-level collective ownership based on production teams would be implemented. In this way, the equalitarianism which reduced the enthusiasm of production teams since the establishment of advanced cooperatives was rectified to a certain extent.

4. Several Issues to Be Further Discussed and the Summary

4.1. *Issues to be further discussed*

a. The natural disaster version does not hold water?

According to the research of Song Guoqing, climate might exert significant effects on the grain output of a local region, whereas the fluctuations of disasters lead to only 1% change of the aggregate grain output of a country because the good and bad harvests in different regions could almost offset one another.

In fact, the analysis in this chapter was carried out based on official statistical materials, and it is impossible to know if the analysis is true or not. Scholars hold different opinions regarding the influence of natural disasters. Nevertheless, one thing is certain, even if natural disasters did lead to crop failures, they could not possibly have caused the largest famine in history. The so-called "Three Years of Natural Disasters" claimed by the government was a sheer excuse.

b. Research on the Death Toll in the Rural Areas

c. Explanation for the Differences in Death Tolls of Different Regions

For example, why were the death rates of Sichuan, Anhui and Henan high while those of Hubei and Fujian were low? Differences existed among the death rates of different provinces. An intuitive impression is that: The death rates of regions which executed the orders of the upper authorities more strictly, such as, accomplishing the assignments of grain procurement, are higher than those of regions which executed the orders less strictly. The more completely a bad system was implemented, the greater the losses would be. In the final analysis, it was the system which went wrong.

4.1.4. At that time, the domestic market and the foreign market did exist. However, China rejected international aid, whereas the food prices at the domestic black markets were too high to be affordable by ordinary people. For example, it was mentioned in the *Records of Jiangsu Province: Record of Prices* that the food price of the black market in 1962 was 13.53 times that of 1960.

> *China did not admit food shortage to the international commu-*
> *nity until the beginning of 1961. At that time, the international*
> *community had already prepared to provide aids, and even the*
> *Taiwan Government had offered 100,000 tons of grains to help the*
> *mainland. However, Foreign Minister Chen Yi told the Japanese*
> *guests that China would never "beg the U.S. for aids". The Chinese*
> *government sternly and contemptuously rejected all aids; it also*
> *refused to accept the aids offered in sincerity by some neutral*
> *international organizations, such as, the International Federation*
> *of Red Cross. The Red Cross Society of China sent a telegram to*
> *the International Federation of Red Cross in Geneva, saying that:*
> *"No famine occurred in China".*

e. Based on the various aspects studied previously on food security, special attention could be placed on research of food availability within a country as well as food availability for households and individuals.

At the national level. During the famine, especially in 1959 and 1960, China refused to accept any international aid and did not import food because Mao Zedong considered import as revisionism and giving the glad eye to capitalism. To make things worse, grain output dropped dramatically due to insufficient labor force in rural areas which was caused by the Great Leap Forward.

At the individual level. In the first place, no large number of people were starved to death in the cities (this point of view is to be further verified, but it is widely accepted) mainly because of the urban-biased system of Unified Procurement and Marketing of grain and the strategy of lowering the urban ration standard. Under the condition that food shortage existed throughout the country, the ration standards in urban and rural areas were both lowered (surely, the rural rational standard was much lower). On the other hand, if the urban-biased system of Unified Procurement and Marketing of grains did not exist at that time and if grains were allowed to flow freely in the country, grain prices would be extremely high, and given the income levels at the time, it was most probable that large-scale famine would have occurred even in cities. This was a method which sacrificed the rural areas to preserve the urban areas. Hence, our opinion is that the strictly executed urban-biased system of Unified Procurement and Marketing of grain helped to preserve the urban areas to a certain extent and that this method is not totally unreasonable. Regarding the rural areas, the opinion of Amartya Sen is totally understandable. On the one hand, the excessive procurement of grain resulted in insufficient grain reserves; on the other hand, what was even more important, the distribution mechanism had been completed disrupted. Peasants were not allowed to travel freely to other regions to beg for food due to restrictions on free flow of population.[2] The lack of food flow mechanism and the restrictions on free flow of population together led to the occurrence of the tragedy of famine.

In the final analysis, the three years of famine occurred in China mainly because there was something wrong with the food availability for individuals and households.

f. Empirical analysis of various influencing factors.

4.2. *Brief summary of the three years of famine*

Such a large-scale nationwide famine in times of peace could only have been caused by human factors. In the early days of New China, per capita grain output was lower than that of 1957 and 1958. Why was there no nationwide famine at that time? Though natural disasters resulted in certain crop failure, the decline in the grain output of 1959 was not caused mainly by natural disasters. Even if grain output decline occurred, there were still the domestic and international markets in which grains could be traded freely. Besides,

---

[2]On December 18, 1957, the CPC Central Committee and the State Council jointly issued the *Instructions on Prohibiting the Blind Outflow of Rural Population* in which a series of measures were stipulated to strictly prohibit peasants from entering cities, including strict inspections by departments of communications, repatriation of "aimlessly drifting population" by departments of civil affairs, strict control of permanent urban residence certificates by public security organs, prohibition on providing food to the "aimlessly drifting population" by departments related to grains, prohibition on recruiting both permanent and temporary workers by employing units, etc. On February 4, 1959, the CPC Central Committee issued the Instructions *on Prohibiting Flow of Rural Labor Force*. On March 1, 1959, the CPC Central Committee and the State Council jointly issued the *Urgent Circular on Prohibiting the Blind Outflow of Rural Labor Force*.

if the famine victims were allowed to flow freely, starvation could also be avoided. China's rejection to import food from the international market and accept aids was mainly affected by ideology. In domestic markets, a strict urban and rural dualistic structure was created artificially and the state exercised absolute monopoly of the procurement and marketing of grains. Though the policy of unified procurement and marketing of grain aiming to preserve cities helped to avoid the occurrence of famine in urban areas to a certain extent, it resulted in a major catastrophe to rural areas. The so-called communism, including the Great Leap Forward, the Wind of Exaggeration, the excessive procurement of grains which exceeded the survival limits of peasants, the Movement of People's Communes which violated economic laws, "large in size and collective in nature," "the combination of political and social organizations" and "Communal Dining Halls Campaign", ultimately deprived the peasants of their last right to subsistence.

The Great Famine from 1959 to 1961 is an exemplification of a bad political system, at the same time, it also showed that the disasters brought about by artificial interruption of markets and deprivation of freedom are far severer than natural disasters.

Chapter 6

Retrospect and Prospect of China's Food Trade

China is one of the major grain producing and consuming countries. Scholars disproved the assertion of *Who Will Feed China* by Brown from different perspectives. Given the limited resources of China, comparative advantages should be exerted and grain trade structure should be adjusted so as to import land-intensive and water-intensive grains and export labor-intensive grains. However, the situation of practical development is not as stated above. Some scholars demonstrated that land was not one of the contributing factors of insufficient grain production in China from the perspective of free trade of land property rights. On the premise that China had become a net exporter of grain products, some researches showed that the low degree of tradability of grain products restrained the functioning of market mechanism and comparative advantages (Li Xiaozhong, 2005). From the perspective of national food security, the framework of grain trade was proposed as follows: according to common international practice, a country's grain supply should rely mainly on domestic production; when grain self-sufficiency rate remains above 95%, grain problem will be solved and the so-called "Big Country Effect" will not occur. This chapter reviews China's grain trade, analyzes the history and reality of China's grain import and export, discusses the prospect of China's grain trade in the future, and analyzes a series of problems.

1. Overall Changes in China's Grain Trade Volume

China occupies an important position in the world grain market, especially in terms of wheat and rice (Wang Hongguang, 2005). Since the 1950s, China's

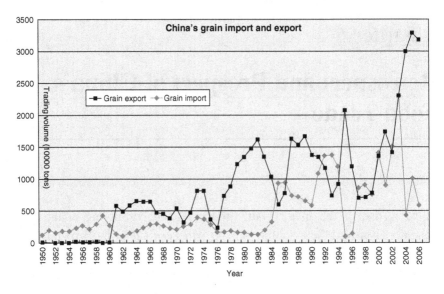

Figure 6.1. China's Grain Import and Export.
Sources: China Customs Statistics Yearbook, Yearbook of China's Foreign Economic Relations and Trade (over the years).

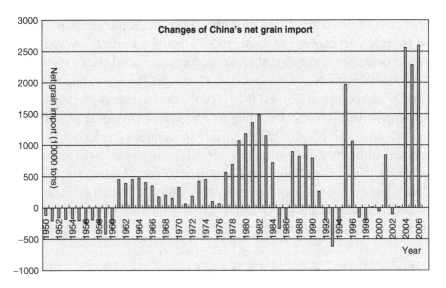

Figure 6.2. Changes in China's Net Grain Import.
Sources: China Rural Statistical Yearbook, 1980–2006; Yearbook of China's Foreign Economic Relations and Trade, 1980–2006.

grain trade volume has been expanding; the trend of net grain import and export show characteristics of frequent changes; the abnormal phenomenon of "buying at high prices and selling at low prices" exists in China's grain trade practice (Lu Jin, 2006). The highest ratio of aggregate grain trade volume (the sum of import and export) to aggregate grain output was 8.5%, whereas the ratio of net grain trade volume to aggregate grain output is basically rather low and the highest net import rate was 5.5% (see Figs. 6.1 and 6.2). According to the nature and history of China's grain trade since 1950 (Qushang, 2004), China's grain trade can be divided into five stages.

1.1. *The 1950s*

In the 1950s, China had been a net exporter of grains and its net export volume showed a tendency to increase year-by-year, rising by 3 million tons or 2.6 times from 1.159 million tons in 1950 to 4.156 million tons in 1959. During this period, China's grain export far exceeded import. Its grain trade basically showed a similar changing trend with that of net trade volume. China's grain import was negligible with the highest import volume being 0.224 million tons which accounted for less than one tenth (merely 7.7%) of the export volume for that year. Besides, in 1959, when the "Three Years of Great Famine" started, China's grain export reached a peak of 4.158 million tons in 1950s, whereas grain import for the same period was almost the least at merely 2,000 tons. In terms of trade volume, the aggregate grain export was 26.7 times that of the aggregate grain import during this period.

The industrialization strategy of promoting agriculture through the development of industry was established in China in the 1950s. In order to serve grain trade in the early days of New China, several sectoral import and export corporations in charge of foreign trade were set up in China. Among them, China National Cereals, Oils and Grainstuffs Import and Export Corporation was the direct operational organization of import and export of grains in New China, in charge of unified management of grain import and export in accordance with the mandatory plans drafted by the State Planning Committee. Planned purchase from the peasants was practised for grains to be exported while planned allocation and sales was carried out for imported grains. Profits and losses are borne by the state financial system. In order to promote the rapid development of industrialization, the state forced down grain procurement prices through the policy of Unified Procurement and Marketing of Grains, sacrificing the interests of the peasants by using "scissors differential", and organized large volume grain exports so as to obtain foreign exchange and to purchase advanced foreign technologies and equipments. At the working conference of the CPC Central Committee in 1961, Chen Yu remarked that foreign exchanges which was acquired through grain export in the past few years was mainly used to purchase complete sets of equipment and important industrial raw materials (Editorial Office of the Grain Work in Contemporary China, Ministry of Commerce, 1989).

As grain prices were artificially forced down in China and foreign exchange was mainly used to promote industrialization at that time, little grains were imported. The aim of importing grains was to exchange needed goods and to regulate supply and demand. Grain import volume during this period was usually very small. Foreign exchange inflow was badly needed in the early stages of industrialization, and net grain export made great contributions in solving this problem. It was estimated that foreign exchange obtained through net grain export made up 12% to 19% of the total foreign exchange obtained through export (Lu Feng, 2004). In 1953, China exported 3.2 billion Jin of grains, among which 2 billion Jin of soybean was exported to the Soviet Union, mainly for exchange of machineries with the Soviet Union and other countries; 0.54 billion Jin to Ceylon (the present Sri Lanka) for rubber and the remainder to other countries. It is clear that the international grain trade in 1950s was mainly about importing grains to regulate supply and demand and exporting grains for foreign exchange and machinery and equipments so as to serve the primitive accumulation of industrial capital (Qu Shang, 2006).

1.2. *The 1960s and 1970s*

The famine from 1959 to 1961 was also a turning period for China in terms of international trade of grains: from a net exporter of grains in the 1950s, it transformed into a net importer of grains after the famine. Starting from the years 1960–1961, grain export was substantially reduced and grain import was greatly increased. Figure 6.3 shows that China's aggregate grain output started to decline in the late 1950s, from 197.69 million tons in 1958 it

Figure 6.3. Changes of China's Grain Output and Net Grain Import in the 1960s and 1970s. *Sources: China Customs Statistics Yearbook, Yearbook of China's Foreign Economic Relations and Trade* (over the years).

decreased to 136.44 million tons in 1961. China's aggregate grain output kept increasing from 1962 to 1979 with the aggregate grain output of 1978 reaching 304.78 million tons. Natural disasters starting from 1959 failed to change China's grain trade policies immediately, on the contrary, the same policies continued for another two years. Nevertheless, net grain export dropped from 4.156 million tons in 1959 to 2.654 million tons in 1960. In 1961, grain trade status changed from net export to net import of 4.455 million tons, the trend of grain trade surplus since the founding of New China was completely changed. China's net grain import status lasted until mid 1980s.

For most of the time during 1960–1970s, China was a net grain importer. From 1961 to 1966, the net import volume was maintained at between 3.5 and 4.5 million tons. In the following 10 years, the net import volume decreased greatly and was basically in a state of continuous fluctuations; the fluctuating range was rather large with the highest net import volume reaching nearly 4.5 million tons and the lowest being merely 0.5 million tons. The ratio of net grain import to the aggregate grain output was around 2% with the highest ratio being 3.2% in 1961. China imported 2000 tons of grains in 1959, 66,000 tons in 1960, 5.81 million tons in 1961, and more than 6 million tons from 1964 to 1966. In the 20 years after 1961, the import volume experienced both increase and decrease, but remained at a high level on the whole. The ratio of grain import volume to the aggregate grain output in China was, however, not very high, being slightly over 3% for 1961 and between 2% to 3% for the other years. In terms of grain export, China exported around 2 million tons of grains for most of the years of the 1960s and 1970s, except for 1959 when more than 4 million tons of grains were exported, and occasionally 3 million tons of grains were exported for some other years. Nevertheless, the ratio of grain export volume to the aggregate grain output in China remained below 5% with the highest ratio being 4.9% in 1974 and the lowest being less than 1%. China's aggregate grain import volume from 1961 to 1978 was 2.4 times that of the aggregate grain export volume.

Though China's grain output nosedived in 1959, the government still procured 64.12 million tons of grains that year, the largest procurement from 1950 to 1981, with 11.55 million tons of grain surplus, also the highest figure from 1950 to 1981. Since trade policies were biased in favour of the demand and supply situation in urban areas, China's net grain export peaked in the years 1958–1960, and amidst the destructive Great Famine, the system of Unified Procurement and Unified Marketing of Grains was given an acid test. The Great Famine exposed the inherent contradictions between the two objectives of the system of Unified Procurement and Unified Marketing of Grain (procuring as much low-price grains as possible from the peasants while leaving the minimum amount of grains needed by the agricultural population in rural areas) as well as its terrible potential consequences. This catastrophe shows that, though certain difficulties might be caused if the government failed to accomplish its procurement target, it would be even more dangerous if the government "successfully" procured excessive amount of grains (Lu Feng, 2004). In the 1960s, influenced by the "Great Leap Forward" and the "Movement of Rural People's Communes," grain productivity was rather low and grain output declined; the demand and supply situation for grains was extremely tense. China transformed itself from a net exporter of grains to a net importer of grains since 1961.

As a consequence of China's grave setbacks in agriculture, macro policies placing agricultural development in the first place began to be formulated, and large-scale grain import policies were also formulated and implemented. China then started to become one of the important members of world grain market. Since early 1960s, when wheat trade between China and Canada first originated, Canada, Australia, the U.S. and Argentina became the largest suppliers for China's grain import. Rice was the main variety of grain exported from China to other countries. Most of the rice (around three-fourths) was exported to neighboring Asian countries and regions, e.g., Hong Kong, Malaysia, Singapore, Sri Lanka, etc.; Cuba and Indonesia were also major importers of Chinese rice. China's grain import volume remained stable in the 1960s and 1970s because at that time decisions regarding international grain trade was centralized and unified instead of decentralized and aimed at meeting actual needs. This policy remained unchanged until the late 1970s. During this period, China's grain trade policies were mainly import-oriented so as to make up the large shortfall in domestic grain supply, and grain import and export was conducted under the trade system of the planned economy. It was estimated that from 1961 to 1965, the total foreign exchange expenditure on net grain import accounted for 13% of the total foreign exchange expenditure on import (Lu Feng, 2004). Meanwhile, speculations were carried out in the international market so as to increase domestic grain supply and reserve, and to earn foreign exchange to support development of socialist industrialization; grains were also exported to support people's revolutions throughout the world (Qu Shang, 2006).

1.3. *The early 1980s to the mid 1990s*

With the implementation of the rural household contract responsibility system and reform and opening-up, China's grain output kept increasing and exceeded 400 million tons for the first time in 1984. China's grain output declined slightly in 1985 and 1986, but it remained higher than 400 million tons after 1987, and starting from 1990 onwards, grain output reached 450 million tons. Nevertheless, China was a net importer in grain trade during this period. The net grain import volume of early 1980s continued to go up on the basis of that of 1960s and 1970s to reach 15 million tons in 1982 and slightly over 10 million tons in 1989. Net grain export appeared occasionally during this period. After which, China entered into a period when net export and net import appeared alternately. Even so, the ratio of the net grain import volume to China's aggregate grain output was still rather low with the highest ratio being 4.1%. China's grain export volume showed an increasing trend, rising from 1.6 million excess tons in 1979 to 10 million tons in 1991 and 14 million tons in 1993. The ratio of grain export volume to the aggregate grain output stood between 2% to 3%. China's grain import exceeded 10 million tons in six successive years in the early 1980s with the highest being 16 million tons; it then experienced a slight decrease but remained above 6 million tons; then it picked up again to stand above 10 million tons. The ratio of grain import volume to China's aggregate grain output remained between 3% to 4%. The ratio of China's total grain import volume to the total export volume during this period dropped from 2.4% of the 1960s and 1970s to 1.9%.

China started the rural economic system reform and opening-up to the outside world in late 1970s. From then onwards, China became more and more involved in world economy and began to be widely and profoundly associated with it. In early 1980s, China carried out grain marketing reform, implemented rural household contract responsibility system, reduced grain procurement amount, improved procurement prices and liberalized the grain market. Propelled by the energy released by the household contract responsibility system and stimulated by the improved grain procurement prices, China's grain output witnessed a continuous and supernormal increase. The phenomenon of "difficulty in selling grains" appeared for the first time in the history of New China, and the situation of net grain import was temporarily reversed in 1985 and 1986. However, as the reform policies gave priorities to urban areas, and the phenomenon of "difficulty in selling grains" led to a fall in grain prices, consequently, peasants' income declined and their enthusiasm in grain production inhibited. Besides, in the grain market reform implemented in 1985, the government abolished the system of unified procurement and unified marketing of grains; in its place was a system of contract purchase of grains. The purchase price was a weighted average at a ratio of 30:70—30% based on old base price and 70% based on old surplus price. This pricing system failed to provide the peasants with sufficient incentives (Lu Feng, 2004). After fulfilling the state contract purchase, peasants could sell the surplus grains in the free market. The state also promised to purchase grains without limitations when the negotiated prices in the free market were lower than the old unified procurement price. This, however, led to the reduction in grain output and the grain purchase targets could not be accomplished. The grain market price began to pick up in the second half of 1985 but was still slightly lower than that of 1984. The periodic oversupply of grains in China came to a transition period, causing increase in grain import and decrease in grain export. In the following years, i.e., from 1987 to 1991, China became a big net importer of grains again. Starting from 1985, China's grain output continued to slip and a lingering situation emerged for as long as five years. From 1987 until 1992, China resumed the tradition of large-scale grain import again. In 1992, due to the double-digit inflation, peasants speculated that grain prices would rise and deliberately delayed the delivery of grains thereby aggravated the supply and demand situation and further pushed up grain prices (price hike of up to 60% was witnessed for several months). In 1993, the state implemented the policy of "guaranteeing grain procurement target while liberalizing the grain purchase prices," that is, the contract purchase amount of grains was fixed, whereas the purchase price fluctuated in line with market situation. As a result, China's grain import volume increased in the following years and reached its peak in 1995, while the grain export volume fell to the trough in 1995.

Prior to the 1980s, China's grain trade was first decided by senior government officials and then implemented by the branches of China National Cereals, Oils and Grainstuffs Import and Export Corporation in different regions. After the 1980s, the Cereals, Oils and Grainstuffs Import and Export Corporations in all provinces, municipalities and autonomous regions became independent operators. Before the reform and opening-up, China's grain foreign trade was totally monopolized by the government. The import and export plan was formulated by the government, foreign trade departments implemented policies on behalf of the state, and professional foreign trade corporations carried out import and export transactions. China National Cereals, Oils and Grainstuffs Import and Export Corporation was in charge of grain trade. In comparison with other industries and other products, the degree of monopoly in grain trade was even higher. Two pertinent facts were relevant in this regard: first, though foreign trade of most commodities was liberalized in late 1970s, grain trade remained monopolized by state-owned trade corporations before China's entry into the WTO; second, top government institutions were involved in the formulation of grain import and export plans.

Decision-making process of grain trade. There are few records on the policy formulation process regarding grain import. Through interviews with relevant government officials and researchers, the general decision-making process in the late 1980s can be discerned. Though the departments involved in the decision-making process and certain details may vary during different periods, there was certain stability regarding the basic principles involved. The first stage was preparation of draft plans for grain trade. This work was usually undertaken by the Overall Planning Division, Grain Bureau of the Ministry of Commerce. The draft plans prepared were based mainly on domestic grain balance sheets. Foreign trade departments also participated in the preparation of draft plans by providing information on world market situation. The second stage was submission of draft plans to the State Planning Committee which would examine the feasibility and soundness of the plan from macro economic perspectives. Two issues were given particular attention: (1) The State Planning Committee assessed the implications of the grain import plan on foreign exchange allocation, since there was a competitive demand for foreign exchange among different departments. (2) The plan was also examined in the context of the state budget as grain imports usually involve subsidies from the state. At the third stage, the State Planning Committee, the Ministry of Commerce and the Ministry of Foreign Trade jointly presented the preliminary plan to the State Council for final examination and approval. Alternatively, the plan might be discussed and final decision might be made at the National Grain Work Conference which was usually held at the end of each year. After being approved by the State Council, the final decision on grain imports would be sent to the Ministry of Foreign Trade and Economic Cooperation, which would assign China National Cereals, Oils and Grainstuffs Import and Export Corporation to implement the plans.

Certain changes occurred during early 1990s. For example, in order to break the monopoly by China Cereals, Oils and Grainstuffs Import and Export Corporation, the quota distribution system was introduced. Other stated-owned enterprises with permissions could also engage in grain trade. Also, with the objective of playing a positive role in stabilizing the grain market in mind, the central government set up the State Bureau of Grain

Reserve in 1991, which actively assumed the role of formulating grain trade policies. It is clear that several basic features existed in the decision-making process. The first is the annual grain planning framework. The second is the high level of monopoly. The third is the direct participation by governmental high-level departments in the plan formulation process.

In late 1970s and early 1980s, with the implementation of economic reform, grain output went up substantially and grain trade fluctuations were mostly caused by the supply and demand imbalances in state grain departments. In order to reduce the peasants' burden and to promote grain production, the state cut down the original compulsory grain procurement quotas. Due to lack of prior judgments about the effects of economic reform on growth, grain shortage of state grain departments kept expanding, and it was necessary to make up this shortage through grain import. In 1980, the Chinese government signed agreements with several major grain exporting countries to ensure a grain supply of more than 10 million tons each year before 1984. Grain fluctuations in the mid 1980s could be explained as follows. The introduction of market mechanism into grain sectors resulted in enlarged macro economic fluctuations, whereas grain circulation and trade were still intervened by the government. In addition, the cyclical fluctuations in grain supply and demand would be magnified by the "super regulation effect" of government intervention. Since the government is committed to the function of economic interventions, in times of surplus of grain output resulting in increased state grain stock, it would increase grain export so as to reduce the stock; in times of shortage, it would increase grain import to cope with reduced grain stock and public pressure. As a result, grain trade fluctuations were magnified, and sometimes even the market trends were reversed. Moreover, time-lag effects existed in trade regulation by the state and indirectly enlarged fluctuations in grain supply and demand, thus leading to the deepened fluctuations in grain trade and frequent changes of market directions (Lu Feng, 2004).

Since the reform and opening-up, China made use of international resources and markets to regulate domestic grain supply and demand situation. However, the annual grain import and export volumes were set by the State Planning Committee basing on domestic grain production and supply/demand situation, and then allocated to different provinces and municipalities; enterprises with grain import and export operation right would then sign contracts with foreign suppliers. In other words, the government planned and regulated grain trade through issuing of "licenses". This sort of international grain trade system led to long lag time and low efficiency in decision-making, and grain import and export often showed the same tendency as that of rich or poor harvests. Therefore, regulating effects obviously lagged behind. The resonance between grain import and export with domestic poor and good grain harvests caused sharp fluctuations in China's international grain trade. That is why policies could not effectively play the role of regulating domestic grain demand and supply (Qu Shang, 2006).

Grain trade at this time made up for the shortage in China's grain production and further improved people's living conditions. "Difficulty in selling grains" stimulated industrial structure adjustment, stabilized and developed rural economy, promoted the formation of domestic grain prices, and supported the national industrialization construction and social development through earning foreign exchange.

1.4. *The Mid-1990s and China's Accession to WTO*

During this period, China's aggregate grain output continued to rise and reached 500 million tons for the first time in 1996. Stepping into the 21st century, China's aggregate grain output declined and fluctuated, being merely 450 million tons in 2001, nearly 60 million tons lower than that of 1998. Net grain import volume went up and down with large fluctuations, decreasing from nearly 20 million tons in 1995 to 1.5 million tons in 1997 and 2 million tons in 1998, and then rising to 8 million excess tons in 2001. The ratio of net grain import to China's aggregate grain output was rather low except for 1995 when it reached the 4% level. China's grain export fluctuations were also very large, ranging from merely 1 million tons in 1995 to 14 million excess tons in 2000. The ratio of grain export to China's aggregate grain output was no higher than 3%. China's grain import also experienced large fluctuations during this period, reaching 20 million tons in 1995, decreasing to less than 10 million tons from 1997 to 1999 and picking up to 17 million excess tons in 2001. The ratio of grain import to China's aggregate grain output ranged from 2% to 4% and never exceeded 5%.

The large fluctuations in China's grain import and export during this period testified the economic development law of international trade: from repressing comparative advantages to exerting comparative advantages. In 1995, the system of provincial governors assuming responsibility for the "rice bag" was implemented, China's insufficient land resources were excessively utilized, which repressed comparative advantages and violated economic development laws. As a result, net grain export emerged during this period and the peasants' income continued to decrease. Domestic supply of grains outstripped the demand for grains after 1997. Under the grain trade system of limited registration and quota management, China started to regulate domestic grain supply and demand by making use of the international market, reduced or basically stopped import of grain products whose domestic supply outstripped the domestic demand and raised import of grain products which were highly demanded in domestic market, thus increasing Chinese people's options of high-quality grain products and satisfying their demand for diversified lifestyle. During the grain marketing reform in 1998, the "three policies and one reform" program was carried out. The program included the following main contents: The state-owned grain purchasing and storing enterprises became the sole legal purchasers of peasants' grains and the sole intermediate link between grain producers and the circulation outlets; the state-owned grain purchasing and storing enterprises purchased peasants' surplus grains at the protection prices without any limitations on the amount, and sold grains at reasonable prices; enclosed operation of the grain purchasing fund; acceleration of the reform of state-owned grain enterprises. However, due to the continuous increase of China's grain output, domestic supply of grains outstripped domestic demand for grains again, and "difficulty in selling grains" occurred for the third time. Peasants' income kept declining after 1998 and peasants' enthusiasm for grain production was seriously harmed, which also provided a pre-condition for the peasants to carry out further industrial structure adjustment. During the grain market reform in 1999, the scope of protective purchase price was narrowed, grain varieties enjoying protection purchase price were reduced, and the contract purchase amount and protective price level were both lowered. It is clear that China's international grain trade at this stage placed emphasis on realizing the market value of grains produced by the peasants and on increasing the peasants' income.

1.5. *After accession to WTO (World Trade Organization)*

After Accession to WTO, China's aggregate grain output continued to stand at 450 million tons, the same level as that at the end of the 20th century. However, it fell to a low ebb of 430 million tons in 2003. From 2004 to 2007, "*No. 1 Documents*", aimed at settling the "Problems Concerning Agriculture, Rural Areas and Farmers," were successively issued. Later, China's grain output stepped out of the low ebb and reached nearly 500 million tons in 2006, almost 50 million tons higher than that of the first year after accession to WTO. Apart from increasing the aggregate grain output, China's macro grain policies also changed the grain trade situation. China had a net grain export of 1 million ton in the first year after joining WTO. This situation was transformed to a net grain import of 88,000 tons in 2003, and the net grain import volume passed through 25 million tons by 2006, accounting for 5.2% of China's aggregate grain output. In the year 2003, due to a large decrease in the sown area of grains as well as occurrence of natural disasters including floods, high temperatures, etc., domestic grain purchase price rose substantially and resulted in the rising prices of cereals, oils and grains throughout the country. As a result, China's export of cereals, oils and grains faced a rare grim situation; the grain export declined year by year from 22.79 million tons in 2003 to 5.8 million tons in 2006, which was a 74.6% decrease; the ratio of grain export to China's aggregate grain output fell from 5.2% in 2003 to 1.2% in 2006. On the contrary, grain import showed a strong rising trend, increasing by 2.3 times from 14 million tons in 2002 to 32 million tons in 2006; the ratio of grain import to China's aggregate grain output reached 6.8%. The total grain import volume of the five years after accession to WTO was 2.3 times that of the total grain import volume.

In the first year after accession to WTO, grain import did not increase as expected; on the contrary, grain export experienced a slight rise instead. After accession to WTO, China continued to implement the grain marketing reform which was started from the years 2001 to 2003; grain circulation was liberalized in most provinces, municipalities and autonomous regions, and grain production and circulation both were rather stable. At that time, substantial grain output decrease occurred in major grain producing and exporting countries, soaring international grain price deprived imported grains of their price advantage. Moreover, with abundant grain stock, the increased market supply of grains forced down the domestic grain price, which was favorable to grain export and repressed grain import.

China's grain trade policies comprised mainly domestic support, market access and export competition. In terms of domestic support to agricultural products, China promised a *de minimis* provision of 8.5% when she acceded to WTO. The current support level in China is merely 0.6%, far lower than the promised level. China abolished agricultural tax in an all-round way in 2006, two years ahead of its five-year schedule. Three subsidies were increased from 0.1 billion RMB in 2002 to more than 30 billion RMB in 2006. Grain minimum purchasing price policy was implemented in major grain-producing regions. In order to preserve peasants' enthusiasm for grain production in major grain-producing regions, grain minimum purchasing price policy was continued in these regions, with rice and wheat as the main targets. Although the implementation of minimum purchasing price violates WTO rules, the amount involved was far lower than China's de minimis level. This policy gives the peasants a signal that stable grain prices can be expected; it arouses their enthusiasm for grain production and protects their interests. China integrated agricultural insurance into the system of support and protection for agriculture and put forward the "three subsidies" policy: i.e. subsidies for rural households, insurance enterprises and agricultural reinsurance. The main contents of the "three subsidies" policy includes the following measures: central and local authorities will provide proportional subsidies to rural households according to their agricultural insurance varieties; to provide appropriate subsidies for the operating expenses of policy-oriented agricultural insurance enterprises; to provide financial support to establish agricultural reinsurance system. Through its direct support to the development of agricultural insurance, the government indirectly offers policy support and protections of interests to local agriculture and local households. In terms of trade policies regarding grain import, China implements tariff quota management on staple agricultural products including grains (wheat, corn, rice) and single-tariff management on soybean, horticultural products, animal products, and other agricultural products. In 2004, the weighted-average tariff rate for agricultural products was 8%, far lower than the rate promised on accession to WTO. China had promised to abolish all export subsidies on joining

WTO, which would greatly promote grain import. China became a net importer of grains again in 2004, which might be a turning point indicating that China was about to step into the grain import era (Yu Jingzhong, 2005). Since the end of 2001, the Chinese government adopted two other major measures to promote grain export. First, railway construction fund was abolished: transportation of paddy, wheat, rice, wheat flour, corn and soybean by railways is fully exempt from railway construction fund. Second, tax rebates for export: export of rice, wheat and corn enjoys a zero rate of VAT and is exempt from the output tax.

2. The Variety Structure and Features of China's Grain Trade

Great changes have happened to the structure of China's grain trade. The import and export volumes of wheat, corn, soybean and rice changed continually, and the major import or export varieties also changed at different stages, reflecting the changes in their positions.

2.1. *Wheat*

Figures 6.4 and 6.5 below show that there were large fluctuations in wheat trade volumes, with a large volume of wheat import and a small volume of wheat export. Wheat import showed a declining trend, whereas wheat export showed a rising trend with a low growth rate. China had never exported wheat to the world grain market prior to 1987 and started to export a small volume of wheat in 1988, less than 10,000 tons, which accounted for less than 1% of China's grain import volume that year. This situation of annual wheat export of less than 10,000 tons lasted until 2000. The total wheat export volume prior to the year 2000 was 130,000 tons, out of which, 87,000 tons of wheat was exported in 1993, corresponding to the strong export tendency for that year; the rest of which equaling to 46,000 tons was exported in the other years. China started to export a small volume of wheat in the early 1990s, and prior to 2001, wheat export was mainly for the purpose of adjusting the domestic grain stock structure. The exported wheat was mostly of low quality and basically was meant for feedstuff. With the increase of domestic wheat output and a steady

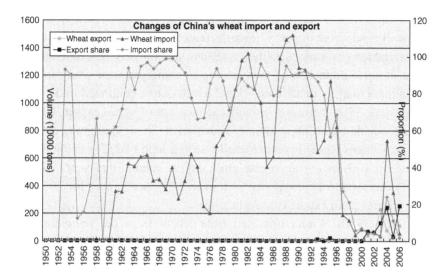

Figure 6.4. Changes of China's Wheat Import and Export.
Sources: *China Customs Statistics Yearbook, Yearbook of China's Foreign Economic Relations and Trade* (over the years).

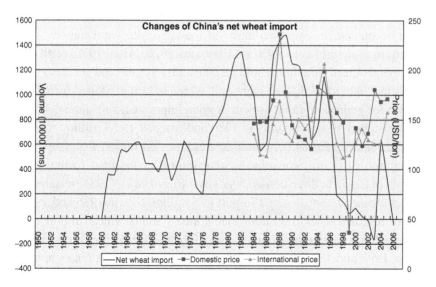

Figure 6.5. Changes of China's Net Wheat Import.
Sources: *China Customs Statistics Yearbook, Yearbook of China's Foreign Economic Relations and Trade* (over the years).

decline of domestic wheat consumption, domestic wheat stock went up continually and wheat import volume decreased continually. In 2002, major wheat-producing countries suffered serious natural disasters and wheat output dropped dramatically, thus leading to rising prices of wheat in the international market; in addition, global wheat output continued to decline in 2003 and 2004; therefore, through the increase of wheat export, China helped to revitalize the international wheat market (Shang Qiangmin, 2005). In 2003, China's wheat export reached a record high of 2.23 million tons, making up 10% of the aggregate grain export volume that year; wheat export decreased to 1 million excess tons in 2006, accounting for 20% of the aggregate grain export volume that year.

In the 1950s, China imported little wheat from the international market, usually less than 0.1 million tons each year, except in 1958 when 0.148 million tons of wheat was imported, which made up two thirds of the aggregate grain import volume that year. Due to three years of natural disasters, China became a net importer of grains from 1961, with wheat as the main imported grain variety. From the beginning of the 1960s, China's annual wheat import increased continually and rapidly rose to 3.58 million tons in 1961. It then stabilized within the range of 3 to 6 million tons, which accounted for over two-thirds of the aggregate grain import. This proportion came close to 99% in 1969 and 1970. After 1978, with the implementation of rural household responsibility system and various types of reforms, individual income increased, which in turn generated increased demand for grains and rapid growth in grain import. China's annual wheat import reached 10 million tons in 1980 and reached 13.53 million tons in 1982, accounting for 84.0% of the aggregate grain import that year. The acceleration of grain circulation system reform contributed to the good harvests in China in three consecutive years from 1982 to 1984, resulted in rapid decrease of annual wheat import to 5 million tons in 1985 and 1986. History repeated itself soon after that. China's annual wheat import rose to 10 million excess tons from 1987 to 1992 and fell to 6 million excess tons in 1993 and 1994. Influenced by policy adjustments, China's annual wheat import bounced back to 11.58 million tons in 1995. Due to the rapid development of high-quality wheat and serious overstock, China's annual wheat import dropped to 1.86 million tons in 1997 and less than 0.5 million tons in 1999. Stepping into the 21st century, China's annual wheat import fluctuated, being less than 0.45 million tons in 2003. By 2004, due to the fact that demand was more than supply for a few consecutive years since 2000, wheat stock continued to decrease, and the state had to liberalize import quotas so as to maintain the balance of supply and demand. A policy of zero-rate VAT for wheat import was also implemented, which greatly reduced the costs of wheat import. China's annual wheat import went up to 7.23 million tons in 2004 and then declined to merely 0.58 million tons in 2006.

China had been a net importer of wheat. After accession to WTO, the scenario of the alternative occurrence of net wheat import and net wheat export emerged gradually. For most of the time of the 1950s, China's annual net wheat import was less than 50,000 tons. Net wheat import rose rapidly to 3.58 million tons in 1961 and 6.21 million tons in 1966; it was maintained within the range of 3 to 7 million tons until 1979 and broke through 10 million tons in 1980 and then decreased slightly in 1985 and 1986 to stand at about 10 million tons. From 1997 onwards till 2001, net wheat import continued to decrease till a level of below 0.5 million tons was reached. The year 2002 was a turning point when China's status as a net importer of grains was transformed into that of a net exporter. Historically, wheat import had gone up mainly because the varieties and quality of part of domestic wheat products failed to satisfy domestic market needs. After implementing adjustment in wheat production structures for many years, low-quality wheat in the northeast and the south decreased substantially while high-quality wheat for special use was promoted in major wheat producing regions, which greatly changed the situation of shortage of high-quality wheat for special use. The development of high-quality wheat varieties led to improved commodity rate and increased market supply of domestic wheat which replaced part of the imported wheat, thus changing the situation of long-term large-scale net import and reducing China's net wheat import volume substantially year by year. However, a net wheat import of 6 million excess tons reoccurred in 2004. After that, China's net wheat import fell again to reach 0.53 million tons in 2006.

2.2. Corn

China's status as a net importer of corn was transformed into that of a net exporter as the net corn export volume rose continually (see Figs. 6.6 and 6.7). Prior to 1955, China did not have any international

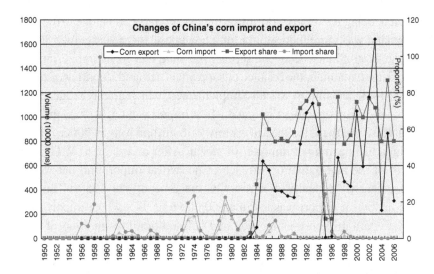

Figure 6.6.   Changes in China's Corn Import and Export.
Sources: *China Customs Statistics Yearbook, Yearbook of China's Foreign Economic Relations and Trade* (over the years).

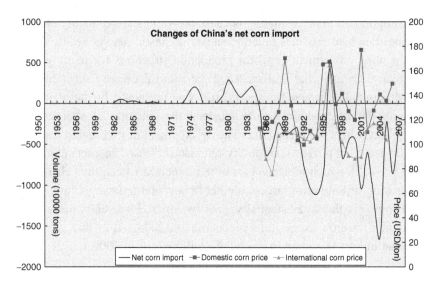

Figure 6.7.   Changes in China's Net Corn Import.
Sources: *China Customs Statistics Yearbook, Yearbook of China's Foreign Economic Relations and Trade* (over the years).

corn trade. China's annual net corn import started to rise from 1956 and remained below 0.1 million tons before 1961; it reached 0.49 million tons in 1962 and remained below 0.3 million tons in the following ten years before 1972. Annual net corn import broke through 1 million tons level for the first time and reached 1.6 million tons in 1973 and 1.9 million tons in 1974; it fell to below 10,000 tons in 1977, but from 1979 to 1983, it was maintained at above 1.5 million tons. 1984 saw the turning point of China's corn trade when China became a net exporter, and scored a net export volume of 0.89 million tons that year. China remained a net exporter of corn in the following ten years with the annual net export volume standing at above 3 million tons. In 1993, China's annual net corn export broke through 10 million tons; it fell to 5.06 million tons in 1995 and 0.28 million tons in 1996. From 1997 until 2011, China remained a net exporter of corn with large fluctuations in net corn export volumes. China's annual net corn export volume was more than16 million tons in 2003 but fell to merely 2 million excess tons in 2004.

China's corn export grew out of nothing and the export volume increased continually. The ratio of corn export to the aggregate grain export continued to rise and reached 86.5% in 2005. China had no corn export before 1982 and started to export corn for the first time in 1983 with an export volume of 60,000 tons. Large amounts of Chinese corn entered the international market in 1985 and the export volume reached 6.33 million tons, accounting for 68% of the aggregate grain export volume that year. China's annual corn export volume decreased slightly after that but did not go below 3 million tons; it broke through 10 million tons for the first time in 1992, accounting for 81.3% of the aggregate grain export volume that year. Soon after that, China's domestic corn price soared and the state adjusted corn import and export policies by replacing corn export with import. Therefore, China's annual corn export volume dropped abruptly to 0.1 million excess tons in 1995 and 1996, which accounted for around 10% of the aggregate grain export volume for those years. After 1997, China's annual corn export restored gradually with an export volume of 6.61 million tons in 1997, and its proportion in the aggregate grain export volume rose to 77.5%. Starting from 1997, supply and demand for corn gradually changed from shortage to surplus in supply. In particular, after 1998 when corn output hit a record high, the huge corn stock led to the continual decline in corn prices, and China exported nearly 5 million tons of corn in 1998. China started to encourage corn export and reduce the sown areas of corn from 1999. Consequently, corn stock gradually decreased and corn prices picked up gradually. With high subsidies from the government, China's corn export volume broke through 10 million tons reaching 10.47 million tons in 2000, an all time high for that time. Domestic corn prices went up again in 2001, and corn export volume decreased dramatically in comparison with that of the same period in 2000. China exported so much corn in these two years mainly because the government provided a subsidy of as high as 44 dollars per ton, which rendered Chinese corn with a competitive edge in the international market. The significant increase in corn export in 2002 could be attributed to the soaring corn prices in the international market and the effects of implementing the policy of exempting staple cereals from railway construction funds and the policy of zero-rate VAT for corn export, both policies helped to lower export quotations and enhanced the competitiveness of Chinese corn. China's annual corn export volume reached an all-time high of 16.4 million tons in 2003, and the market readjusted when a turning point in supply and demand emerged because corn stock had declined to a rather low level. In 2004 and 2005, supported by state policies, China's domestic corn output rose continually, whereas corn prices picked up steadily because domestic demand for corn showed a strong increase. In 2005, China's corn export volume reached a new peak of 8.64 million tons, and the corn export volume accounted for 86.5% of the aggregate grain export volume.

China started corn import long ago, but the corn import volume was quite low and accounted for a small proportion of the aggregate grain import. China did not import any corn prior to 1955; it started to import corn in 1956 with an annual import volume of 12,000 tons. China's annual corn import volume was below 0.1 million tons before 1961; it stood at below 0.2 million tons for most of the years from 1962 to 1972 except in 1962 when 0.49 million tons of corn was imported, which accounted for 10% of China's aggregate grain import. Corn import exceeded 1 million tons for the first time to reach 1.6 million tons in 1973; it then dropped dramatically from 1974 onwards and reached a mere 8,000 tons in 1977. Corn import soared to 2.79 million tons in 1979, after which fluctuated greatly, importing only 1,000 tons in 1991 while no corn was imported in both 1992 and 1993.

From 1994 to 1996, grain shortage forced domestic grain prices to go up, and China's grain price experienced a rare steep rise, as a result, China imported 5.18 million tons of corn in 1995, which was a record high. After 1996, corn import never surpassed 0.5 million tons; it did not exceed 0.1 million tons after 1999, and no corn was imported from 2003 to 2005. The good harvest of corn in China in 2004 effectively increased domestic market supply and augmented the declining stock.

2.3. *Rice*

Rice is one of China's major export grain varieties and once occupied an important position in the international grain market, with a stable annual export volume of less than 3 million tons (see Figs. 6.8 and 6.9). China started rice export in 1950 and its annual rice export volume reached 1.67 million tons in 1956, accounting for 63% of the aggregate grain export that year; rice export volume reached 1.77 million tons in 1959 and then decreased gradually; it did not exceed 1 million tons from 1961 to 1965, accounting for around 40% of the aggregate grain export. Rice export volume stood at above 1 million tons for almost all of the fifteen years from 1966 to 1980; it reached 1.73 million tons in 1973, accounting for about 50% of the aggregate grain export that year, and this proportion reached the peak at 76% in 1978. Although political struggles and the ultra-Left ideology prevailed during the Cultural Revolution, however, pragmatic policies were adopted in grain trade. In the international grain market, the price of rice was normally twice that of wheat although the two cereals contains similar amount of calories per physical unit. A trade model of exchanging rice for wheat was developed by making use of the price advantage of rice over wheat (Lu Feng, 2004). During this period, China remained a net grain importer in terms of physical volumes; however, a balance was maintained in terms of trade value. China's annual rice export volume declined slightly after 1981, standing at around 0.5 million tons, which accounted for 30% to 45% of the aggregate grain export, for three consecutive years. From 1985 to 1997, rice export volumes stood at below 1 million tons for most of the years, with the lowest being merely 67,000 tons, accounting for 6.5% of the aggregate grain export; rice export volumes exceeded 1 million tons for five years, reaching at 1.77 million tons in 1994, accounting for 15% of the

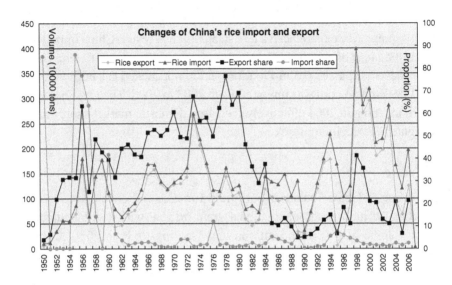

Figure 6.8.   Changes of China's Rice Import and Export.
Sources: *China Customs Statistics Yearbook, Yearbook of China's Foreign Economic Relations and Trade* (over the years).

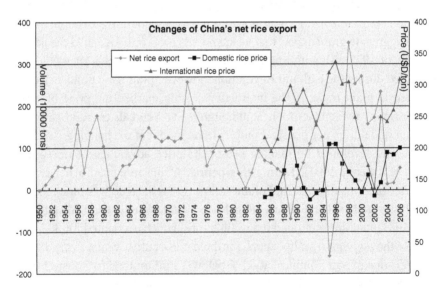

Figure 6.9.   Changes of China's Net Rice Export.
Sources: *China Customs Statistics Yearbook, Yearbook of China's Foreign Economic Relations and Trade* (over the years).

aggregate grain export that year. Rice export reached a record of 3.74 million tons in 1998, accounting for 41% of the aggregate grain export that year; this proportion then started to fall to merely 6.9% in 2005. The reasons for the decline in China's rice export volume after 1998 are as follows: the broken rate of Chinese rice was high and the overall quality was relatively low, the price fetched in the international market was significantly lower than the domestic market, thus the overall competitiveness of Chinese rice was not strong in comparison with the rice from other major rice exporting countries, such as, Thailand, Vietnam, the U.S., etc. (Liu Lifeng *et al.*, 2006).

China's annual rice import volume basically remained below 0.5 million tons. The proportion of rice import volume in China's aggregate grain import volume was quite high before the 1960s and then dropped to below 10%. China imported 57,000 tons of rice in 1950, accounting for 85.4% of the aggregate grain import volume that year. China stopped importing rice from 1951 to 1954. The annual rice import volume then stood at around 0.1 million excess tons in the following three years, accounting for 60% to 80% of the aggregate grain import volume; it exceeded 0.3 million tons in 1961, accounting for 6% of the aggregate grain import volume that year. The proportion of rice import volume in aggregate grain import volume remained less than 10% from the 1960s to 2011, ranging from 1% to 5% most of the time. China's annual rice import volume stabilized at around 0.1 to 0.2 million tons from 1962 to 1982. From 1982 to 2011, rice import volume stabilized at below 0.5 million tons for most of the years, except in 1989 and 1995, when it reached 1.01 million tons and 1.62 million tons respectively. China's rice import volume quota was as high as 3.99 million tons in 2002, far higher than the actual annual rice import volume for the other years. The great expansion of rice import volume quota had a significant influence on rice import. China's annual rice import volume stood at 0.236 million tons in 2002, slightly higher than that of 2001. From 2003 onwards, there was an oversupply of rice in domestic market, and Chinese rice encountered poor sales and serious overstocking. Although Thai fragrant rice met the demand of some medium-high income residents in big and medium cities in the southeast coastal areas of China for improved quality, its price was more than three times higher than that of the high-quality japonica rice in northeast China (Ya Zhou, 2003) and common people do not take it for their daily meals. These two factors limit import volume of Thai rice.

China has basically been a net exporter of rice since 1951. China's annual net rice export volume was less than 10,000 tons in 1950; it remained below 2 million tons for most of the years before 1997 except in 1972 when 2.56 million tons of rice was exported. In 1989, a net rice import volume of 0.67 million ton was spotted which changed China's long-term position as a net exporter of rice. China's annual net rice export volume reached 1.6 million excess tons in 1994. Due to extended shortage in domestic grain supply and low rice quality, rice export volume declined continually, and its proportion in the international rice trade dropped continually. Net rice import appeared in two consecutive years from 1995 to 1996. Annual net rice export volume exceeded 3 million tons and reached a new record of 3.5 million tons in 1998. From 1998 to 2003, annual net rice export volume was maintained at between 1.5 and 2.5 million tons. Starting from 2004, due to the low quality of Chinese rice, China's rice export gradually declined and the annual net rice export volume stood at below 0.5 million tons.

2.4. *Soybean*

China started to export soybean in 1959 with an export volume of 0.91 million tons, accounting for 74.5% of its aggregate gain export volume that year. Annual soybean export volume broke through 1 million tons in 1951 and then saw a slight decline in the following few years; it remained above 1 million tons from 1955 to 1960, accounting for 40% to 50% of the aggregate grain export volume. Soybean export volume stood at below 0.7 million tons from 1961 to 1983 and showed a declining trend; it dropped to 0.11 million tons in 1978, accounting for merely 6% of the aggregate grain export volume that year; it remained above 1 million tons in the remaining years of the 1980s and in the early 1990s, accounting for 10% to 20% of the aggregate grain export volume; it showed a long-term decline from 1992 to 2011, standing at 0.17 million tons in 1998, accounting for less than 2% of the aggregate grain export volume that year. In the first five years of the 21st century, it fluctuated between 0.2 and 0.3 million tons; in recent years, it did not increase significantly, standing at merely 0.4 million tons, accounting for 6.6% of the aggregate grain export volume.

Over the years, China's position as a net exporter of soybean was changed to that of a net importer. Its annual soybean export volume remained at around 1 million tons, whereas annual soybean import volume reached 28 million tons in 2006, accounting for 89% of China's aggregate grain import volume that year. China imported little soybean prior to 1972, except for 1952 to 1954, when 1 million tons were imported annually, and no soybean was imported in other years. For more than 20 years after 1973, China's annual soybean import volume stood at below 1 million tons, with the highest being 0.71 million tons, accounting for 8.8% of aggregate grain import volume that year; for most of the years, it stood at below 0.5 million tons and even at below 0.1 million tons for nine years. Starting from 1996, in order to meet the rapidly increasing domestic demand for soybean, China temporarily abolished the soybean import quota policy and lowered the import tariff from 114% to 3%. China's annual soybean import volume broke through 1 million tons in 1996; it accelerated with a soaring growth rate for the following four consecutive years, surpassing 2 million tons in 1997, 3 million tons in 1998, 4 million tons in 1999, and jumped up to 10.41 million tons in 2000. In the protocol for accession to WTO, China formally promised to abolish the quota management on soybean import and to implement a single tariff of 13% of VAT. The implementation of this policy actually opened up China's soybean market completely and led to a drastic increase in China's soybean import volume. Starting from 2003, China's annual soybean import volume remained at above 20 million tons and reached 28 million tons in 2006, which made up 88.9% of China's aggregate grain import volume for that year.

China was mainly a net exporter of soybean before 1995 with net import emerged in only a few years. Net soybean import started in 1996 and the annual net soybean import volume grew rapidly. Since there was no soybean import before 1972, China's annual export volumes and annual net export volumes of soybean showed an identical trend during this period (please see Figs. 6.10 and 6.11). Net soybean import stood at around 1 million tons before 1960; it reached the peak at 1.72 million tons in 1972 and was maintained at below 0.7 million tons from 1961 to 1973. Net soybean import occurred for the first time in 1974, with a net annual soybean import volume of 0.24 million tons, however, it was replaced by net soybean export in the following two years. Net soybean import appeared again from 1984 to 1995, with the net annual soybean import volume lingering at around 1 million tons for seven consecutive years. Starting from 1996, China entered the period

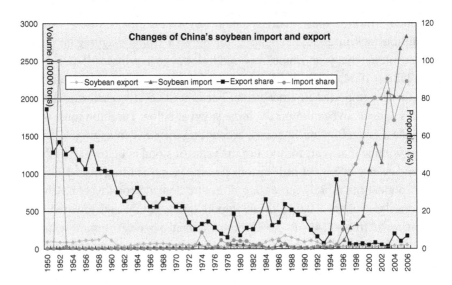

Figure 6.10. Changes in China's Soybean Import and Export.
Sources: *China Customs Statistics Yearbook, Yearbook of China's Foreign Economic Relations and Trade* (over the years).

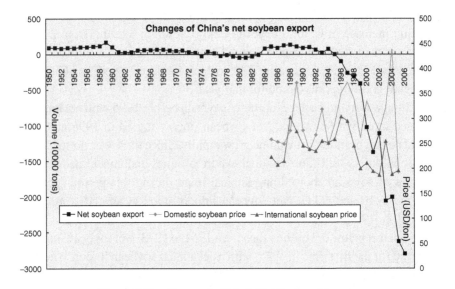

Figure 6.11. Changes in China's Net Soybean Export.
Sources: *China Customs Statistics Yearbook, Yearbook of China's Foreign Economic Relations and Trade* (over the years).

of net soybean import and the annual net soybean import volume continued to rise from less than 1 million tons in 1996 to 28 million excess tons in 2006.

2.5. *The overall structural changes*

The objectives of structural adjustment of grain varieties were changed from implementation of trade policies to economic efficiency, from an overall balance of quantities with improved grain varieties to making use of market mechanism to improve resource allocation efficiency, promote grain and agricultural industrial structure adjustment and to exert the comparative advantage and improve the efficiency of grain production. China's increased domestic demand for grains, especially grains used for the transformation of food structure, e.g., grains used for animal feed and wheat of inferior quality, was met primarily by the international market and the cultivated land resources of other countries. Obviously, there is certain economic rationality in such policies.

2.5.1. *Changes of grain import*

Though great changes have happened to the structure of China's grain import varieties since 1950, China's grain import still centered on one variety only (see Fig. 6.12). Prior to 1960, the structure of China's grain import varieties kept changing, with soybean as the main grain import variety in 1951, wheat in 1953 and 1957, and rice in 1954 and 1956. From 1960 to 1996, wheat was the main grain import variety of China, accounting for 60% to 99% of the aggregate grain import volumes. Starting from 1997, soybean became the main grain import variety, and its proportion in the aggregate grain import volume rose from 45% in 1998 to 88.9% in 2006. Wheat import witnessed slight increases occasionally.

2.5.2. *Changes of grain export*

It can be seen from the above Fig. 6.13 that China's grain export varieties have been changing all the time, with several grain varieties occupying the

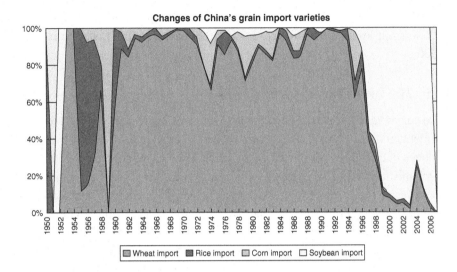

Figure 6.12. Changes of China's Grain Import Varieties.
Sources: *China Customs Statistics Yearbook, Yearbook of China's Foreign Economic Relations and Trade* (over the years).

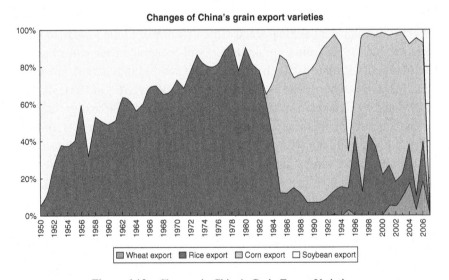

Figure 6.13. Changes in China's Grain Export Varieties.
Sources: *China Customs Statistics Yearbook, Yearbook of China's Foreign Economic Relations and Trade* (over the years).

prime positions. Prior to 1982, China's grain export comprised rice and soybean with few other varieties; out of which, rice accounted for 40% to 60% and soybean 20% to 40% of China's aggregate grain export volume. In the 1980s, corn became China's major grain export variety, while soybean and rice also occupied large proportions; out of which, soybean made up 10% to 20%, rice around 10%, and corn about 50%. Recently, corn still occupies a prime position in grain export; soybean export decreases continually; rice export and wheat export both show an increasing trend. In 2005, wheat export volume accounted for 2.6% of the aggregate grain export volume; soybean accounted for 4%; rice accounted for 6.9%; whereas corn accounted for 86.5%.

3. Food Embargo and Food Self-Sufficient Rate

As China's grain import kept increasing, China's food self-sufficiency rate went down continually and its foreign trade dependence degree kept rising. As a large country, China is facing an increasing risk in terms of food trade, especially the threat of "food embargo" in international sanctions. Food embargo is a political issue which deserves our attention in the future development of China. Many scholars' research on the issue of food embargo concluded that the probability of a food embargo imposed on China was rather low (Lu Feng, 1997). This research indicates that food embargo on China may be probable, but if food self-sufficiency rates are maintained within a certain range, China will be able to cope with economic embargos and to ensure food supply.

3.1. *A review of international food embargos*

Food embargo generally refers to the reduction in the amount or even discontinuation of food circulation in the international market caused by policy interventions. In the history of world food trade, large food exporting countries reduced or even discontinued their food supply to food importing countries for the sake of their political goals, sometimes solely or mainly in the form of food trade and sometimes in the form of a package of embargos. Food embargo has four aims: to "punish" or "take revenge on" the food importing countries; to safeguard the economic interests and political stability of food exporting countries; to interfere in the internal affairs of and affect the political situation of food importing countries; to force the recipient countries to adjust their policies in accordance with the wishes of food exporting countries through restricting or stopping the so-called "concession export" of food which took on the nature of aids to these countries (Hu Yingchun, 2003).

The research of Hufbauer and Schott (1985) suggested that 103 cases of economic sanctions occurred throughout the world from 1919 to 1984. Among the 84 cases of economic sanctions that occurred from 1950 to 1985, 10 cases involved food embargos. Among the 10 cases of food embargos, nine were targeted at developing countries and one was targeted at the Soviet Union; eight cases were initiated by the U.S. (Winters, 1990) and one case was China's imposition of food embargo on Vietnam during the Sino-Vietnamese War; five food embargos totally failed, two caused short-term pressure on embargo targets but with no long-term influence, one succeeded yet with negligible effect, and two basically succeeded (Winters, 1990, p. 176).

At least half of the 10 food embargos belonged to partial or complete discontinuation of food export for aid purposes rather than commercial embargos on food export. The remaining are mostly food embargos in the broad sense, i.e., embargos targeted at large numbers of trade items including food. Only one belongs to food embargo in the narrow sense as defined above. The two successful food embargos had a common feature, i.e., their targets were recipients of food as aids. The reasons for the failure of food embargos are as follows: the interchangeability and substitutability of food; divergence of interests among food exporting countries; and humanitarian considerations. Food embargo is a double-edged sword, in the WTO era, and imposing food embargo on a country also causes economic loss of major food producing countries.

3.2. Analysis of the possibility of food embargo on China

After WTO accession, the degree of dependence of China on international food trade continued to rise. In this section, we will make an analysis on the possibility of a food embargo against China, taking into consideration the possible causes and consequences of food embargo, as well as the national conditions of China.

3.2.1. *Possibility of food embargo*

According to researches on food embargos around the world, though negligible, the possibility of a food embargo against China does exist. A number of scholars believe that food was usually excluded from the embargo list in the past for humanitarian reasons, and the probability of occurrence of food embargo is not high under current social conditions. However, it is possible that an economic sanction against China may cause the reduction of food supply.

The United Nations Security Council adopted Resolution 661 and took steps to impose mandatory economic sanctions and weapon embargos on Iraq. It decided that all states shall prevent the sale or supply of any commodities or products, including weapons or any other military equipment, except for medical or humanitarian purposes, to any person or body in Iraq or Kuwait, and that all states shall prevent the availability of funds or other financial or economic resources to the government of Iraq.

The United Nations Security Council adopted Resolution 666 which regulated principles for supplying foodstuffs to Iraq or Kuwait. It decided that the Security Council Committee shall keep the situation regarding foodstuffs in Iraq and Kuwait under constant review, and requested the Secretary-General to urgently seek information on the availability of food in Iraq and Kuwait, communicating all information to the Security Council Committee. The resolution also decided that if a humanitarian situation arises to supply foodstuffs to Iraq or Kuwait in order to relieve human suffering, the Council would meet to discuss it immediately, directing the Committee to direct foodstuffs through international organizations such as the International Committee of the Red Cross and the United Nations, and to distribute by them or under their supervision in order to ensure that they reach the intended beneficiaries.

The United Nations Security Council then adopted **Resolution 670** which decided to enforce an air blockade on Iraq and the occupied Kuwait. The resolution decided that Member States should deny permission of aircraft to take off from their territory if it should take cargo to or from

Iraq and occupied Kuwait, excluding medical and humanitarian aid; and that Member States should deny permission to any aircraft destined for Iraq or Kuwait to overfly its territory.

The imposition of economic embargos on Iraq by the UN from 1996 to 2003 seriously affected food supply. Iraq was prohibited from selling oil to other countries, which deprived Iraq of sufficient economic resources; at the same time, it was prohibited from purchasing items from the international market. Despite of the humanitarian assistance (the United Nations Security Council Resolution 1284 called upon the Government of Iraq to take all steps to ensure the timely and equitable distribution of all humanitarian goods, in particular medical supplies), wide-reaching famine occurred in Iraq within a few years. As a result, a large number of children suffered from malnutrition, and some people were starved to death. In order to alleviate the suffering brought about to Iraqi people by sanctions, the UNSC adopted Resolution 986 in April 1995 (namely, the Oil for Food Agreement) which allowed Iraq to export $2 billion worth of oil every half year for importing essential foodstuffs and medicine. From the account above, it can be seen that embargos on Iraq involved many contents. In addition, it was clearly decided that all states should prevent the transportation of any commodities, except medicine and foodstuffs for humanitarian purposes, to any person or body in Iraq or Kuwait, and that foodstuffs and medicine should be provided through and distributed by the International Committee of the Red Cross which was authorized by the UN. That is to say, no commercial food trade existed.

In UN's resolution regarding sanctions on North Korea, clauses related to food included the prevention of the supply of luxury goods to the Democratic People's Republic of Korea (DPRK); in the sanctions on Iran, clauses relevant to economy included keeping alert to Iran's financial activities. These sanctions had not reached the scale of economic sanctions.

The UN's Resolution on Imposition of Sanctions on DPRK: The United Nations Security Council adopted the Resolution 1718 regarding the test of a nuclear weapon by the Democratic People's Republic of Korea in the afternoon of 14 October 2006. According to reports, sanctions in this resolution mainly included six aspects: (1). The resolution expressed the gravest concern at the claim by the Democratic People's Republic of Korea (DPRK) that it had conducted a test of a nuclear weapon. (2). The resolution was passed under Chapter VII, Article 41, of the UN Charter. (3). All Member States should prevent the direct or indirect supply, sale or transfer to the DPRK of all items, materials, equipment, goods and technology which could contribute to DPRK's nuclear-related, ballistic missile-related or other weapons of mass destruction-related programs; member States were banned from exporting luxury goods to DPRK. (4). All Member States should freeze immediately the funds, other financial assets and economic resources of individuals and companies involved in or providing support for DPRK's nuclear-related, other weapons of mass destruction-related and ballistic missile-related programs. (5). Shipments of cargo going to and from North Korea might be stopped and inspected for weapons of mass destruction or associated items. (6). Underlined that further decisions would be required, should additional measures be necessary.

It can be seen that similar circumstances might also exist in China. In other words, the possibility of food embargos on China also exists. Firstly, some or a few countries might impose food embargo on China. This sort of embargo will not affect China because China could obtain food from other countries or solve the problem of food supply by itself. Secondly, international organizations, including the UN, and their member states might impose embargos on China. If this kind of all-round embargos is imposed, China will be unable to obtain foreign exchange or other income from the international market, and its links to the outside world will be cut off. Under such circumstances, China will not be able to obtain financial resources from the outside world or import food. Besides, the possibility of food embargos caused by joint economic sanctions imposed on China by several countries also exists.

3.2.2. *Strategies to solve the problem of food embargo*

In case where economic sanction is carried out jointly by several countries on China, it might affect the availability of foreign resources, but it will not greatly affect food supply. Instead, countries which carry out food embargos will suffer immense losses of foreign trade. For instance, food sanctions imposed on the Soviet Union by the U.S. caused the U.S. to suffer a loss of $2.2 billion. When China's food self-sufficiency rate remains above 95%, China could acquire food supply from other countries in the world. If China's food self-sufficiency rate stands above 90%, apart from international trade,

China could also make use of food stocks to maintain a food supply for two months. During this period, the rising food prices will drive the increase of investment in food planting in China, including input of human resources and capital, etc., thus guaranteeing the increase of food supply within two months.

If an imposition of joint blockade is initiated by the United Nations on China, this will pose certain challenges to China's food supply, but it is still within the capability of China to dissolve it. Comparing with Iraq, economic sanctions on China will lead to different results. Iraq depended heavily on food import, economic sanctions thus led to severe social unrest. Sanctions on Iran and North Korea did not include food embargos and the supply of foodstuffs subject to authorization by the UNSC is allowed under humanitarian circumstances. The case of Iraq will not happen to China because China's food self-sufficiency rate generally will not fall below 90% and basically remained above 95%. Under current food trade conditions, maintaining the normal functioning of granaries for two months will be able meet the short-term urgent need. Besides, China does not rely on import for cereals, soybean, which is the grain variety with the largest annual import volume, is mainly used to produce edible oil. If a food embargo is imposed on China, cereals supply will not be greatly affected and it will not be a problem to prevent its people from suffering starvation. Regarding soybean, on one hand, China could meet the demand by using stocks; on the other hand, China could promote the increase of investment in soybean planting and improve the supply of soybean within a quarter.

The possibility that a food embargo may happen, undoubtedly poses a test to China's food self-sufficiency rate. If the food self-sufficiency rate stands at above 95%, China would not be affected greatly; if it falls between 90% and 95%, China will be able to get through the difficulty within a short period of time; if it falls below 90%, China will pay a heavy price to cope with the crisis.

Therefore, the impact of a possible food embargo on China still exists, but China will be able to overcome the crisis very quickly. Nevertheless, China has to ensure that its food sufficiency rate stands above 95% so as to guarantee China's domestic stability and to fulfill its responsibility as a large country.

3.3. *Analysis of the self-sufficiency rate of 95%*

The article *Who Will Feed China* published by the American scholar Brown in 1994 brought about a heated debate on China's food security. The national food security index is a standard for assessing food security. FAO set up three basic standards: the first standard is that a country must try its best to raise its food self-sufficiency rate to above 95% (security in food production); the second standard is that per capita food consumption should reach 400 kilograms (security in food consumption); the third is that food reserve (food stock) should reach 18% of the food consumption that year, and that 14% is the minimum level, food reserve below which will lead to a state of food emergency (security in food circulation). In *China's Food Security Report* submitted to the World Food Summit in Rome by the Chinese government in November 1996, it was pointed out that, in order to guarantee the national food supply security, China's food self-sufficiency rate will not be lower than 95%. Many years of experiences have proved that tight food supply and large food price fluctuations will occur in China when its food self- sufficiency rate falls below 95%. Therefore, China should take the food self-sufficiency rate of 95% as the baseline of China's food security, and its net food import volume should be kept below 5% of the aggregate domestic food consumption. The food self-sufficiency rate of 95%, which has been used by China for quite a long time, is both a food self-sufficiency rate essential for China's food security and a promise made by China as a responsible large country. The origin of the food self- sufficiency rate of 95% is as follows:

First, since China possesses certain food production capacity, 95% of self-sufficiency rate could technically be guaranteed. China's current annual food consumption is approximately 480 billion kilograms, and its current annual food demand grows at about 1% and is unlikely to accelerate. Since China's annual food production capacity has come close to or reached 500 billion kilograms, it is rather easy for China to maintain a stable annual food production of 450 billion excess kilograms. Even if food stock is insufficient, the balance of food demand and supply will be maintained through an import of 25 to 30 billion kilograms of food, which accounts for 6% to 7% of China's aggregate annual food consumption. This import volume is safe and will not pose large threat to food security.

Second, proper food import is economical in the long run. High social and economic prices will be paid for the pursuit of excessively high food self-sufficiency rate. Theoretically speaking, the larger the food import, the more beneficial it is to China's environmental protection and resource utilization. China has a large population but limited land resources with scarce cultivated land and water resources, and has no comparative advantage in land-intensive food. The growing demand for land-intensive food is inflexible, whereas food output increase brought about by technology is limited. In addition the opportunity costs of cultivated land and other resources are going up. Under such a macro context, the pursuit of excessively high food self-sufficiency rate will inevitably intensify the problem of shortage of scarce resources, and the reoccurrence of destruction of ecological environment will be inevitable.

Third, the international market has the potential of increasing food supply. Countries in Asia, Latin America and Africa with serious food problems have launched Green Revolution and established the system of food reserve one after another. Countries relying on food import, e.g., Japan, South Korea, etc, have formulated the policy of improving food self-sufficiency rate. After realizing self-sufficiency of staple food, the EU vigorously promoted agricultural protection policies so as to consolidate its position of self-sufficiency of staple food. According to the experiences of China and the world, the adverse weather will cause food output to fluctuate by 1% to 2%. The global food reserve generally exceeds 10% of the annual food consumption of the world. Advances in agricultural science and technology provide a basis for China to import a suitable volume of food. In addition, the increase of per unit area yield and the expansion of harvest areas will raise the aggregate food output of the world to at least 10.4 billion tons in future. Therefore, famine will not reoccur in the future world. Besides, China has certain purchasing power and the self-sufficiency rate of 90% is already very high in the world. Thus, for China to import a suitable volume of food is essential for the development of world economy and to ensure the integration of China's domestic market with the international market.

Meanwhile, China must stick to the principle that food supply should basically be based on domestic resources, and to implement the policy of importing and exporting food at the same time. After China's accession to the WTO, according to the minimum market access rules, China's food import volume should not be lower than 5% of its domestic demand for food, and the net food import volume should not be higher than 5% of its domestic food supply. China has a large population but limited land resources. With seven percent of the world's cultivated land, China has to feed one fifth of the world's population. Once China enlarges food import, great influence will be exerted on the international food price. In view of this, the international community keeps a vigilant eye on China's food import volume. Generally speaking, China's food security is rather secured. Two key information will make this clear: China's food self-sufficiency rate has been stabilized at over 95% for more than a decade, the balance in food production and demand is basically realized; China's state food reserve was increased by 15 billion excess kilograms in 2006 when compare with 2005, the stock-to-sales ratio far exceeded the minimum security range of 17–18% set by the FAO.

First, large country effects determine that China must keep a foothold at home. China's food output and consumption both constitute around 21% of the aggregate food output and consumption of the world, and China's food trade volume accounts for 15% to 20% of the trade volume of the world. If China's food self-sufficiency rate declines by 1%, China will need to import 5 billion kilograms of food, which makes up 2.5% of the trade volume of the world. If China depends heavily on international resources and market for realization of its food security, China will constitute a threat to food security of the world. After China's accession to the WTO, China raised the upper limit of food import quota to 21 million excess tons in 2004, which equaled to 4% or 5% of China's current annual food consumption and apparently would not cause substantial harm to food security.

Second, the interaction and interweaving of the price effect and income effect of food determines that China's food self-sufficient rate must be around 95%. In the long term, food prices in China are significantly higher than the international food prices, and a certain pattern of food security protection actually has been formed, i.e., China hardly exports any food, whereas food import mechanism is activated. On top of this, the comparative advantage of some labor-intensive foodstuffs in China might lead to increase in export, which will exert positive income effects, i.e., export increase leads to higher income, which in turn leads to more food consumption and food import. Nevertheless, due to large country effects, the increase of food import will drive the international food prices up, and the rise of food prices will in turn stimulate domestic food production, thus restraining the increase of food import. Due to the effect of this interaction mechanism, it is very hard for China's food self-sufficiency rate to fall to below 70% to 80% even under free trade conditions. In reality, traditional trade barriers and domestic support and protection still exist in developed countries, and new trade barriers are being reinforced. Neither economic growth nor the improvement of food security could be solved by depending solely on trade liberalization. Therefore, China's food self-sufficiency rate should not be lower than 90%, or better still, should remain around 95%.

Third, from the perspective of institutional economics, institutional changes will effectuate increasing returns and self-reinforcement mechanism. The idea of attaching great importance to food security has a special historical origin in China. An old Chinese saying "food is deemed paramount by the people" reflects the central idea that political rulers must not allow the people to face starvation. The Chinese government has set the lower limit of food self-sufficiency rate at 95%. The case of food self-sufficiency rate being lower than this limit is considered rather impossible to happen.

4. Evaluation of the Prospects of China's Grain Trade

Based on an analysis of the history of China's grain trade and an evaluation of China's current trade policies and the circumstances under which such policies are implemented, we will make an assessment of the overall trend of China's grain trade in the near future and put forward our ideas on how to guarantee China's grain security and to exert China's comparative advantages at the same time.

4.1. *A net importer of grains*

Time series are used here to make predictions. Since too many influencing factors would be involved if the time span is too long, therefore a time span of ten years is selected to predict China's aggregate grain import and export. Generally speaking, China's future grain trade will be dominated by net import and the annual net grain import volume will continue to rise and reach approximately 35 million tons in 2010, in which the aggregate grain export volume will reach 11 million excess tons, and the aggregate grain import volume will be 46 million excess tons. By 2020, China's annual net grain import volume will reach 65 million tons, in which the aggregate grain export volume will reach 13 million excess tons, and the aggregate grain import volume will be 78 million excess tons.

Meanwhile, both domestic and foreign experts and institutions also made predications on China's grain trade situation. For instance, researchers from University of California, Davis and International Food Policy Research Institute (IFPRI) believed that, China's annual grain import volume will be 28 million tons in 2010 and 30 million tons in 2020; if China's population continues to grow at a low rate of 1.4%, China's annual grain import volume will 56 million tons in 2020. Fan Shenggen and M.A. Sombilla believed that the baseline forecast value of China's grain import volume in 2020 is 41 million tons; if the government does not increase investment in agriculture, China's grain import volume in 2020 will reach 85 million tons.

Following is an assessment of the trend of development of China's grain trade. First, judging from China's trade situation, China's net grain import keeps increasing in recent years, and the continuation of net grain import is understandable. Second, China's increased demand for grains, industrial raw materials, etc, will cause China to consume more imported commodities. Thirdly, due to the deepening agricultural reform in China, rural land productivity is about to be saturated, limited increase can be expected, whereas demand will increase continually, grain import is thus needed to make up the gap. Last but not the least, due to changes of China's grain policies and the application of comparative advantages, China will continue to exchange land-intensive or resource-intensive products with labor-intensive goods; since grain products generally belong to land-intensive products, China will increase its demand for grain import and continue to be a net importer.

4.2. *Forecasting on trade situations of different varieties of grain*

4.2.1. *Wheat*

Judging from the overall development trend of wheat, China's wheat export volume will go up slightly, but it will stand at around 1 million tons in the short run. China's wheat import volume, however, will remain rather unstable and fluctuate greatly, the highest may be above 5 million tons and the lowest may be less than 0.5 million tons. Therefore, China's net wheat import volume will also show certain volatility and both net import and net export will occur. The highest net wheat import volume may be 3 million excess tons, whereas the net wheat export volume will remain around 1 million tons.

4.2.2. *Corn*

China's corn trade was dominated by import prior to 1983; starting from 1984, corn trade was dominated by net export. In recent years, China remains a net exporter with the annual export volume standing at around 10 million tons, and corn is the major grain export variety of China. China's annual corn import volume had been relatively small, standing at below 0.1 million tons. Therefore, it had little influence on China's aggregate grain import.

4.2.3. *Rice*

For more than five decades, China's annual rice export volume stood at below 1.5 million tons for most of the years, with the highest being 3.75 million tons; China's largest annual rice import volume was merely 1.63 million tons. Only around 1% of China's domestic rice output participated in international trade, and the proportion of China's rice export volume in international rice export volume never exceeded 10%. Therefore, the position of rice trade is likely to fall. Judging from historical data of rice, annual rice export volume may stand at around 1 million tons in the near future, and the proportion of rice export volume in the aggregate grain export volume will decrease slightly to around 10% due to adjustments of trade policies; annual rice import volume may remain at below 1 million tons or may rise up slightly. On the whole, net rice export will continue in the near future, but a period of net rice import will emerge and the net rice import volume will increase rapidly after its first appearance.

4.2.4. *Soybean*

China's soybean export has shown a declining trend, its annual export volume may remain at below 0.5 million tons, accounting for around 5% of China's annual aggregate export volume. On the other hand, annual import volume of soybean shows a strong growing trend and may break through 30 million tons soon, accounting for about 90% of China's annual aggregate import volume. Therefore, soybean will become China's major grain import variety in the future, and net soybean import will remain in the long run.

Chapter 7

China's Grain Distribution

1. Background

1.1. *Multi-objective grain policies*

According to Chen Xiwen (2002) and Du Ying (2002), China's grain policies have the following objectives: Grain security, stable grain price and supply, reduction of fiscal burden and income increase of grain farmers.

The game of interests exists between the central government and local governments, among different regions and between the central government and state-owned grain enterprises.

1.2. *The feature of grain production*

When reviewing China's grain production from 1980 to date, we will find certain periodicity. A possible explanation is as follows: The opportunity cost of agricultural production increases when macro economy grows at a high rate, shifting the demand curve to the left; then grain demand increases and high inflation comes along with rapid economic growth, leading to the farmers' reluctance to sell grain and speculative grain storage; when economic recession occurs, the reverse applies. Theoretically, this sort of periodic fluctuations could be smoothed out by international trade and grain stock.

Lu Feng put forward the hypothesis that China's grain policy relaxes when economic situation is easy and tightens when economic situation is tough which corresponds with production cycles so as to explain the fluctuations in the reform of grain distribution field. The inconsistency of grain policies further magnifies the periodic phenomena. After 1998, a consensus that grain supply and demand would need to be maintained in a tight state seemed to have been formed. Therefore, China's grain policy was tightened when the relationship between grain supply and demand was rather easy in the grain reform in 1998.

2. The Process of China's Grain Distribution Management System Reform

The policy changes in China's grain distribution management system since the foundation of new China could be roughly divided into the following stages:

2.1. *The stage of free purchase and marketing of grain (1949–1952)*

Salient features: Free purchase and marketing of grain, market-determined grain prices and diversified business entities.

2.2. *The stage of unified procurement and marketing of grain (1953–1984)*

Salient features: unified procurement and marketing of grain, state-determined grain prices and monopolized management by the state-owned grain system.

2.3. *The stage of double-track grain procurement and marketing system (1985–1998)*

In 1985, the unified procurement and distribution of grain was abolished and contract procurement was implemented; grain outside the contract procurement was allowed to be freely traded in the market. By the end of 1993, grain price and purchase and marketing had been basically liberalized in 98% of the counties (cities) of China. During the inflation period in 1994, grain procurement was resumed and the prices of major grain varieties were determined by the state. The central government clearly put forward the system of provincial governors assuming responsibility for the rice bag in 1995. According to Grain Procurement Regulations implemented in 1998, the state procured grain without limitations, monopolized grain sources, sold grain at market prices and implemented the enclosed operation of grain procurement funds. At the National Grain Procurement and Marketing Work Television Conference in 1998, the State Council proposed that the state would procure grain without limitations at protective prices and that the state-owned grain enterprises would sell grain at market prices, generally at 0.04 RMB per kilogram.

2.4. *The stage of free purchase and marketing of grain (1999 to date)*

2.4.1. *Year 1999*

Guo Fa (1999) No. 11 proposed to appropriately narrow the scope of grain procurement at protective prices, to complete the policy of grain procurement prices. This would enable grain prices to be set in accordance with grain quality; improve and complete the means of fiscal subsidy to grain procurement; and permit large agricultural industrialization enterprises and feed enterprises to enter the grain market for grain purchase within provinces after getting provincial-level administrative approval.

2.4.2. *Year 2000*

In view of the difficulty of state-owned grain enterprises in selling grain at market prices and the problem of aged grain, it was decided that the scope of grain procurement at protective prices should be further narrowed and that grain purchasing enterprises should be permitted to purchase grain directly from rural areas, after getting approval from county-level administrative departments. (Guo Fa, 2000, No. 12)

2.4.3. *Year 2001*

Guo Fa (2001) No. 28 proposed "to liberalize grain selling regions, to protect grain producing countries and to enhance regulation with provincial governors taking responsibility". Grain and feed procurement were abolished in eight major grain selling regions, namely, Zhejiang, Shanghai, Guangdong, Jiangsu, Fujian, Beijing, Tianjin and Hainan. Meanwhile, the following were required: (1) basic farmland should be firmly protected and conversion of cultivated land to non-agricultural land without permission should be strictly prohibited, which was an impassible "red line"; (2) provincial grain reserve should be replenished in accordance with the state requirement of "grain selling region maintaining grain stock for grain sales of six months" and the responsibility and capacity of provincial-level people's governments in grain regulation should be enhanced.

By the end of 2003, grain procurement and sales marketization reform had been implemented in Zhejiang, Shanghai, Guangdong, Jiangsu, Fujian, Beijing, Tianjin, Hainan, Guangxi, Yunnan, Chongqing, Qinghai, Guizhou, Anhui, Hunan, Hubei, Inner Mongolia and Xijiang (the latter five regions are major grain producing regions). The government dominated the market price of grain through grain procurement without limitations at the minimum procurement prices in all the other regions.

In terms of the nationwide trend of grain supply and demand, grain varieties which would be procured at the minimum procurement prices in the future would mainly include middle-season and late-season paddy in the middle reaches of the Yangtze River, high-quality paddy in the northeast, wheat in the Huang-huai-hai region and corn in the northeast and the east of Inner Mongolia.

2.4.4. *Year 2003*

Article 18, Article 19 and Article 20 of Regulations on the Administration of Central Grain Reserves permitted the storage of grain reserves of the central people's government on a commission basis.

By the end of 2006, 1773 enterprises qualified for the storage of grain reserves of the central people's government on a commission basis (see Table 7.1).

2.4.4.1. Operation of storage of grain reserve on a commission basis

Article 20 stipulates that enterprises which meet the requirements for storage of grain reserve on a commission basis as referred in Article 19 would qualify for storage of grain reserve of the central people's government on a commission basis after being examined and approved by the state grain administrative departments.

Article 21 stipulates that China Grain Reserves Corporation is responsible for selecting best qualified enterprises for storage of grain reserve

Table 7.1. Classification of Ownership Forms of Enterprises Qualified for the Storage of Grain Reserves of the Central People's Government on a Commission Basis.

Forms of business ownership	Number of enterprises	Proportion (%)	Storage capacity (10,000 tons)	Proportion (%)
Total	1,773		9,201.3	
State-owned or state-controlled	1,769	99.8	9,185	99.8
Private-owned	4	0.2	16.3	0.2
Foreign-funded	0	0	0	0

Source: *China Grain Development Report 2007*

of the central people's government in accordance with the overall layout plan of the grain reserve of the central people's government, reporting to the state grain administrative department, the financial department of the State Council and Agricultural Development Bank of China for records, and sending a duplicate to local grain administrative departments.

2.4.4.2. The procurement and marketing plan of grain reserves
    of the central people's government

Article 14 stipulates that the procurement and marketing plan of grain reserves of the central people's government would be proposed by the state grain administrative department in accordance with the scale, grain varieties and the overall layout plan of grain reserves of the central people's government approved by the State Council. This would be then examined and approved by the development and reform department and the financial department of the State Council and transmitted to China Grain Reserve Corporation by the development and reform department of the State Council and the state grain administrative department jointly with the financial department of the State Council and Agricultural Development Bank of China.

2.4.5. *Year 2004*

2.4.5.1. Regulations on the administration of grain distribution

Article 3 stipulates that the stage encourages market entities of various ownerships to engage in grain management activities and promotes fair competition. Grain management activities according to law shall be protected by the state laws. Illegal means to hinder free grain distribu-tionwould be strictly prohibited.

2.4.5.2. Clarification of the examination and approval requirements for grain procurement

Article 8 stipulates that any dealers engaged in grain procurement activities shall meet the following requirements:

(1) Having the ability to raise the operating funds;
(2) Owning or having the ability to rent necessary facilities for grain storage;
(3) Having corresponding capability of quality testing and safekeeping for grain.

The specific requirements of the preceding paragraph shall be formulated and promulgated by the people's governments of provinces, autonomous regions or municipalities directly under the Central Government.

Article 9 stipulates that the dealers qualified for grain procurement shall be registered pursuant to of the People's Republic of China on Administration of Registration of Companies before they can engage in grain procurement activities.

While applying for engagement in grain procurement activities, dealers shall submit written applications and supporting materials regarding their funds, storage facilities, quality testing and storage abilities, to appropriate grain administrative departments and departments in charge of industrial and commercial registration. Grain administrative departments shall complete the examination and approval process within 15 working days from the date of receiving the application. They will approve applicants who meet the specific requirements referred to in Article 8 and make it known to the public.

2.4.5.3. Grain distribution statistics

Article 13 states that grain purchasers should report relevant information on the amount of grain procurement to grain administrative departments of county-level people's governments of the procurement region at regular intervals.

In the case of grain procurement across provinces, grain purchasers should report relevant information on the amount of grain procured to both the grain administrative departments of county-level people's governments of the procurement regions and the regions where grain purchasers belong at regular intervals.

2.4.5.4. Mandatory requirements on grain stock

Article 20 states that dealers engaged in grain procurement, processing and sales must maintain essential grain stock. The people's governments of provinces, autonomous regions or municipalities directly under the central government shall set the specific standards of the minimum and maximum grain stock when necessary.

2.4.5.5. Policy-related grain utilization through market transactions

Article 26 states that the state should implement the grain reserve system at both the central level and local levels. Grain reserves are used to regulate grain supply and demand, to stabilize the grain market and to cope with severe natural disasters or other emergency situations.

The procurement and sales of policy-related grain utilization shall be conducted openly through grain wholesale markets, in principle, or other means stipulated by the state.

2.4.5.6. Procurement at the minimum procurement price

Article 28 states that when significant changes occur in the relationship between grain supply and demand, the State council could decide to implement the minimum grain procurement prices on key deficient grain varieties in major grain producing regions as necessary so as to ensure market supply of grain and the interests of grain farmers.

When grain price experiences or is likely to experience significant increases, the State Council and people's governments of provinces, autonomous regions and municipalities directly under the central government could take interventional measures pursuant to provisions of Price Law of the People's Republic of China.

2.4.5.7. The abolishment of grain procurement regulations

Article 54 states that these regulations shall come into force upon promulgation. Grain Procurement Regulations issued by the State Council on June 6, 1998, and Punishment Measures for Illegal Actions of Grain Procurement and Sales issued by the State Council on 5th August 1998 shall be repealed at the same time.

Guo Fa (2004) No. 17 proposed the following mechanisms:

- The procurement, sales and updations to the inventory of grain reserves of local governments and the central government shall be openly conducted through standardized wholesale markets in principle.
- 2. Domestic grain varieties, and grain shortage and surplus shall be regulated by making use of the international market in a flexible way and in accordance with the principle of "depending mainly on the domestic market and making use of imports and exports for appropriate regulation".
- 3. The overall balance of grain should be guaranteed. Major grain selling regions and major grain producing regions should form a long-term stable procurement and sales cooperation mechanism on the basis of market mechanism and according to the principle of mutual benefit.
- 4. Local grain reserves should be soundly regulated. It should be ensured that local grain reserves should meet the scale requirements of the state, be marketable and of high quality and should be easily transported and utilized when needed.
- 5. The order of grain market should be standardized. Regional blockade and protection are prohibited. State-owned enterprises should play the role of main channel. The behaviors of grain dealers should be standardized.

2.5. Year 2006

Guo Fa (2006) No. 16 proposed the following mechanisms:

2.5.1. Establishment of market-oriented mechanism

In order to further standardize the relationship between governmental regulation and enterprise operation, the government could entrust qualified grain procurement and sales enterprises with policy-related business and grant subsidies to these enterprises in accordance with standards set according to the need of macro regulation of grain.

The procurement, sales and inventory updating by turns of grain reserves of local governments and the central government shall be conducted in standardized grain wholesales markets in the form of competitive price transactions.

2.5.2. *Cultivation of rural brokers*

Diversified grain market entities should be cultivated, developed and standardized continuously. Various qualified market entities are encouraged to engage in grain procurement, sales and management activities. Rural grain brokers should be cultivated so that they can participate in fair competition and enliven grain distribution. Diversified investment entities should be guided to investment in various grain exchange markets, grain logistics facilities and high-tech grain and oil processing enterprises.

2.5.3. *Summary*

Competition should be gradually liberalized. Nevertheless, state-owned grain procurement and sales enterprises should continue to play the role of main channel, and the government should retain its capacity of regulating grain market.

3. The Current Situation of Grain Market

The Eleventh Five-Year Plan (2006–2010) for the Construction of the Nationwide Grain Market System, proposed to establish by 2010, a unified, open and orderly grain market system with grain procurement markets and retail markets as its basis; grain wholesale markets as its backbone, national grain exchange centers as its driving force and grain futures market as its forerunner which combines trade flow with logistics, traditional transactions with e-business and spots with futures, and has a more reasonable layout, endowed with complete functions and a complete institution and more standardized operation.

3.1. *Grain fair trade markets and supermarkets*

By the end of the Eleventh Five-Year Plan, China had over 80,000 urban and rural fair trade markets (retail).

3.2. *Grain wholesale markets*

3.2.1. *Classification*

a. Grain wholesale markets which came into being spontaneously on the basis of fair trade markets are mostly private-owned.
b. Grain wholesale markets which implement the membership system are mostly government-owned.

Presently, there are 600 grain wholesale markets in excess in China. The total annual trading volume of grain and oil from all grain wholesale markets in China stands at 80 million excess tons (the total annual distribution amounts to around 0.2 billion tons), and the total value amounts to around 123 billion RMB.

There are 477 grain wholesale markets in major grain producing regions with an annual trading volume of 52.2 billion Jin; there are 274 grain wholesale markets in major grain selling regions with an annual trading volume of 37.5 billion Jin (Training Tutorial in the Regulations on the Administration of Grain Distribution).

3.2.2. *Transaction mode*

Negotiated transactions refer to transactions concluded through price negotiations between buyers and sellers in private.

Competitive price transactions are used in policy-related grain utilization and grain reserve conversion in principle.

The consequent problem is that many enterprises conduct transactions in the over-the-counter market rather than in the exchanges. In many cases, the market price of grain is lower than the protective procurement price because the protective grain procurement prices were implemented in grain producing regions while the market price of grain was formed in grain-selling regions. As state-owned enterprises cannot sell grain procured at protective prices at market prices, these grains were sold at prices lower than the protective prices through non-market secret means.

Case: Crazy Flour — the Embarrassment of an Entrepreneur and the Grain Distribution System

Date: 12 December 2006 Author: Shi Yu Source: *First Financial Daily*

The minimum procurement price led to market monopoly by China Grain Reserves Corporation (Sinograin).

On 21st November, according to the report of a local media, the procurement price of wheat in Zhengzhou has risen to 1.80 RMB per kilogram, while the normal price at this time in the previous year was 1.36 RMB per kilogram. A shopkeeper whose last name is Liu in a farmer's market at Weisi Road of Zhengzhou City told the reporter that (because of the rising procurement price of wheat) 10 kilograms (of flour) was sold at 22 RMB three days ago (18th November) and is being sold at 28 RMB now. The price of flour has risen by 0.6 RMB per kilogram in three days.

In accordance with objective laws, market circulation of grain usually accounts for 40% of the aggregate grain output, and the normal grain procurement by Sinograin makes up 40%–60% of the total market circulation of grain. China Zhengzhou Grain Wholesale Market predicts that China's wheat output in this year will be around 0.103–0.105 billion tons (about 100 billion kilograms). In accordance with calculations based on general laws, China's grain procurement basically equals the grain circulation of the whole market. This indicates that farmers have sold more grains to Sinograin than previous years, leading to large amounts of grain stock in Sinograin and the diminishing grain circulation in the market. Flour enterprises certainly could not take it any more.

Li Shang, Vice General Manager of Henan Provincial Grain Exchange and Logistics Market, told the reporter that, "The wheat output of Henan province will ascend substantially and exceed 25 billion kilograms this year (the wheat output of Henan province amounted to 25.7769 million tons or around 25.8 billion kilograms. The specific wheat output of 2006 has not been released by authorities). Sinograin

(Continued)

procured (at the minimum procurement price) 18.1 billion kilograms of grains this year while its grain procurement in previous years (when the minimum procurement price has not been implemented) basically stabilized at 7.5 billion kilograms."

He also said that, "Grain belongs to Sinograin after it arrives at the warehouses of Sinograin. The reserve of auction certainly will not be lower than the sum of the minimum procurement price and storage costs. In addition, Agricultural Development Bank of China takes the final responsibility for grain auction of Sinograin. Surely, the more grains Sinograin procures, the more benefit it will make."

The Management Regulation of Grain Distribution promulgated by the State Council stipulates that "the state implements the grain reserve system at both central and local levels" and that "banks should provide timely grain procurement loans to qualified grain buyers in accordance with relevant regulations of the state".

Li Shang also introduced that "therefore, there are both grain depots directly under the control of Sinograin and those owned by local governments or with private equity participation in a particular region" and that "these grain storage enterprises could conduct settlement with Agricultural Development Bank of China and accept the commission of Sinograin to implement the policy of the minimum grain procurement price as long as they are qualified for bank loans".

Gao Tiesheng, former director general of the State Administration of Grain and former general manager of Sinograin, pointed out that, "According to regulations of the policy of the minimum grain procurement price, the costs and losses incurred for grain procurement should be paid by the state finance, and the storage expense of grain procured at the minimum procurement price should also be paid by the state; grain depots were active in grain procurement and unwilling to see the outgoing-depot of grain procured at the minimum procurement price because these grain would make profits for these enterprises continuously as long as they stayed in depots; the fact that the costs and

(Continued)

losses of state-owned grain enterprises incurred for grain procurement at the minimum procurement price were paid by the state finance led to the unwillingness of large numbers of enterprises to participate in the grain distribution system reform; (and ultimately) most of the input of the state finance was used in grain procurement expenses, and farmers did not really benefit much; in accordance with the experiences of other countries in the implementation of the minimum procurement price, farmers could only get 25% of the fiscal expenditure on the implementation of the minimum procurement price".

3.2.3. *The game of interests*

An insider of the grain system at Henan Province told the reporter that, "for instance, in terms of the design of the grain auction system, the present auction amount and auction time are decided through negotiations among the National Development and Reform Commission, the State Administration of Grain, the Ministry of Finance and Sinograin. The Ministry of Finance once proposed round-the-clock open auction, complete disclosure of information about grain sources and online auction participated by grain purchasing enterprises at any time, so as to reduce costs and avoid the panic of grain purchasing enterprises. This proposal, however, was never carried out".

3.3. *Futures market*

There are three futures exchanges at present. Zhengzhou Commodity Exchange trades futures in wheat, and Dalian Commodity Exchange trades futures in corn, soybean, soybean meal, and soybean oil (see Table 7.2).

No. 2 soybean contract is designed in accordance with the indicator system of oil content with imported genetically modified soybeans as the main subject matter. Domestic soybean is also used for delivery. For example, No. 1 Soybean contract uses domestic non-transgenic soybean as the subject matter.

Table 7.2.　Agricultural Products Futures Transaction Distribution in 2006.
Monetary unit: 10,000 RMB

Items	Turnover	Trading volume
Strong wheat	50,672,045	29,352,476
Hard wheat	84,083	56,104
No. 1 Soybeans	48,736,392	17,794,122
No. 2 Soybeans	9,924,324	3,850,452
Corn	204,000,000	135,000,000
Soybean meal	147,000.000	63,099,338

Heilongjiang Province organized its soybean planting plans by referring to the prices of soybean futures transacted in Dalian Commodity Exchange.

4. Grain Brokers

In recent years, the nation-wide grain circulation grows larger and larger, standing at around 0.2 billion tons. Moreover, State-owned grain enterprises carried out large-scale layoffs. In comparison with the situation of 1998, the total number of state-owned grain enterprises decreased by 25,400 or 47.7% and the total number of grain purchase and sales enterprises reduced by 12,700 or 41.8% in 2005; the total number of staff in the state-owned grain enterprises fell by 2.171 million or 66%, and the total number of staff in the grain purchase and sales enterprises dropped by 1.203 million or 62%. Besides, as tens of millions of young and middle-aged rural labor force went off to the cities for work, the destitute women, children and the elderly found it very difficult to send their grain to grain management stations for sale. In this background, grain brokers emerged as the key players.

Rural grain brokers in China belong to three categories. The first category includes rural grain specialized cooperatives (such as, wheat associates, grain and oil service centers, etc.) which carry out grain purchase. For example, the Jinli Wheat Associates in Yanjin county, Henan province purchased over 50% of the total wheat sold by peasants in 2006. The second category includes peasants with experience of running small businesses and who conduct grain purchase. The third category includes laid-off workers from grain enterprises.

The proportion of the amount of grains purchased by rural grain brokers in the total amount of grain purchased by state-owned grain enterprises is quite high in China, being 60% in some regions and 40%–50% in some other regions.

5. Relevant State Departments of Grain Distribution

5.1. *State administration of grain*

The State Administration of Grain (SAG) is a national administrative agency under the National Development and Reform Commission (NDRC), which is responsible for macro-control of national grain distribution, providing guidelines for the development of grain industry and administrating national grain reserves. The principal functions of the SAG include:

(a) Entrusted by the NDRC, to study and formulate mid-term and long-term strategies of national grain macro-control, overall balance of supply and demand, grain distribution, grain import and export and deployment of national grain reserves; and to develop and implement the programs for the reform of national grain distribution system.

(b) To draft laws and statutes, relevant policies, rules and regulations for the nation's grain distribution and national grain reserves management, and supervise the enforcement; to formulate plans of building facilities for grain distribution, storage and processing, among which large-medium construction projects should be report to higher authorities for examination and approval in accordance with specified procedures; and to raise proposals in regard of setting a framework for grain protective procurement price, protection price and limited market price.

(c) To standardize and manage the quality of grain products in coordination with the State Bureau of Quality and Technology Supervision; and to establish technical criteria for grain storage and transportation, and supervise the implementation.

(d) To steer the management of national grain distribution and personnel training of grain sector; to guide and promote technical reform and the spread of new technology; to promote foreign exchanges and cooperation; and to handle the statistic work of grain industry.

(e) To formulate technical norms for the management of national grain reserves and supervise the implementation; to propose the scale and overall layout of national grain reserves as well as plans for procurement, marketing, import and export of national grain reserves, and supervise the implementation; to supervise and examine the stock, quality and security of national grain reserves. The SAG is also responsible for guiding the business of China Grain Reserves Corporation.

(f) To undertake other tasks assigned by the State Council and the NDRC.

5.2. *China grain reserves corporation*

China Grain Reserves Corporation (Sinograin) was founded in 2000 with a registered capital of RMB 16.68 billion. Entrusted by the State Council, Sinograin's primary mandate is to manage and operate central grain and oil reserves by ensuring the quantity, quality, and storage safety of central grain and oil reserves and accepting the assignment of regulating the purchase, sales, transfer and stock of grain and oil entrusted by the state. Under the macro economic regulation and the supervision and administration by the state, Sinograin operates independently and assumes sole responsibility for its profits or losses. Its main activities include storage, processing, trade and logistics of grain and oil, and research and service of storage technologies.

Chapter 8

International Food Security and Food Trade

Since World War II, the international food security status has greatly improved, with the continuous improvement in comprehensive grain production capacity and people's dietary standards. However, obvious food gap still exists in the world, with food surplus in developed countries and serious food shortage in developing countries. To guarantee food security worldwide, and ensure that people's basic requirement of having plenty to eat could be satisfied, tremendous efforts would need to be taken by all governments and people.

1. World's Food Supply and Demand

1.1. *World's food supply*

Since 1961, the global cereal production showed a continuous rising trend, going up by 2.1 times from 0.88 billion tons to 1.82 billion tons during the rapid growth period of grain output from 1961 to 1985. World's food output went up slightly since 1986 mainly because the developed countries' enthusiasm in food production fell, which led to the decline of food output growth rate and the slow cereal output growth. After entering the late 20th century, though per unit area grain yield increased slightly, world's grain sown area, the total grain output and per capita food output presented a declining trend (see Fig. 8.1).

Certain changes have happened to the cereal layout worldover. The annual global aggregate cereal output stands at around 2 billion tons, among which 1.1 billion tons are produced by the developed countries and 0.9 billion tons by the developing countries. Hereinto, America produces

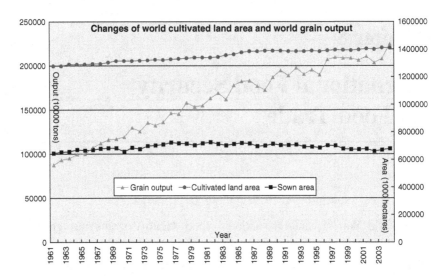

Figure 8.1. Changes in Cultivated Land Area and Grain Output Worldwide.
Source: Data adapted in accordance with data of FAO.

0.5 billion tons of cereals annually with a per capita cereal output of 600 kilograms; Europe produces 0.4 billion tons annually with a per capita cereal output of 560 kilograms; Oceania produces 0.03 billion tons annually with a per capita cereal output of 1000 excess kilograms; with the largest population, Asia produces 1 excess billion tons annually with a per capita cereal output of 330 kilograms which generally equals to the average level of the world; Africa produces 0.15 billion tons with a per capita cereal output of 200 kilograms, the least among all continents.

World's food production exhibits the following features:

1.1.1. *The slow increase in world's food harvested area*

Food harvested area reflects the resource potential for food production. The global food harvested area rose from 0.56 billion hectares in 1990 to 0.61 billion hectares in 2006, with an annual growth rate of 0.5%. The order of annual average growth rate of different regions from top to bottom is as follows: Oceania, Central and South America, African countries south of the Sahara, North Africa and Asia (excluding China), all of whose annual average growth rates stood above the global average level.

The annual average growth rates of the other regions were all lower than the global average, among which China had the lowest growth rate of −0.3%.

1.1.2. *The increase in gross food output promoted by the increase in per unit area yield*

The per unit area yield reflects the overall agricultural production technology level, and its changes represent the growth pace of agricultural technological progress. The increase in per unit area yield is the main driver for the increase in the gross food output, whereas the increase in food sown area only contributes 9% of the increase in gross food output. Therefore, the growth trend in the global per unit area yield basically corresponds with the growth trends of the global gross food output.

The global average per unit area yield was 3.5 tons per hectare in 2006. With the development of science and technology and the improvement of agricultural production conditions, the cereal per unit area yield globally has increased continuously. From 1961 to 2006, the global average per unit area yield rose from 1353 kilograms per hectare to 3.5 tons per hectare, and the global average cereal per unit area yield increased by 1.6 times. Hereinto, regions whose per unit area yield is higher than the global average include Europe (excluding the twelve countries of the former Soviet Union) whose per unit area yield stood at 5.3 tons per hectare, China (5.0 tons per hectare) and North America; the per unit area yield of North Africa and the Central and South America was slightly lower than the global average; the per unit area yield of Oceania was 1.0 ton per hectare, the lowest; the per unit area yield of African countries south of the Sahara was 1.7 tons per hectare. The global average per unit area yield rose from 3.0 tons per hectare in 1990 to 3.5 tons per hectare in 2006 with an annual growth rate of 1.0%. The order of annual average growth rate of different regions from top to bottom is as follows: Central and South America, North Africa, North America, Asia (excluding China) and Europe (excluding the twelve countries of the former Soviet Union), all of whose annual average growth rates stood above the global average level. The annual average growth rates of the other regions were all lower than the global average, among which Oceania had the lowest growth rate of −0.7%.

However, the growth rate of the global cereal per unit area yield showed a declining trend. Generally speaking, the growth in the global cereal per unit area yield can be divided into four stages: 1961–1970, 1971–1980, 1981–1990, and post-1991. The global cereal per unit area yield achieved fastest growth during 1961–1970, with an annual average increase of 53.9 kilograms per hectare and an annual growth rate of 3.0%; it increased by 35.4 kilograms per hectare annually in average during 1971–1980 with an annual growth rate of 1.6% and by 43.7 kilograms per hectare annually in average during 1981–1990 with an annual growth rate of 2.2%; the growth rate of the global cereal per unit area yield declined after 1991 with the annual growth rate stabilizing at 1.2%.

In terms of individual countries, China, Canada and France had the highest growth rates of per unit area yield of cereal, namely, 3.4%, 2.5% and 2.4% respectively; per unit area yield of cereal of Brazil, India, the U.S., Germany was higher than the global average.

In terms of the total amount, the order of the following countries in the per unit area yield of cereal from top to bottom was as follows: Netherlands, France, Germany, Japan, the U.S., China, Argentina, Canada, Brazil, India and Australia, whose average per unit area yield of cereal during 1961–2005 was 6,827.2 kilograms per hectare, 6,946.5 kilograms per hectare, 6,657.5 kilograms per hectare, 6,027.7 kilograms per hectare, 4,714.6 kilograms per hectare, 3,031.1 kilograms per hectare, 2,582 kilograms per hectare, 2,195.3 kilograms per hectare and 1,860 kilograms per hectare respectively.

1.1.3. *The declining growth rate of global cereal production*

Grain output is the result of both harvested area and per unit area yield. The aggregate global cereal output stood at merely 0.88 billion tons in 1961, around 2 billion tons during 2000–2004 and at 2.14 billion tons in 2006, with an increase of 144% and an annual average growth rate of 2.0%. Hereinto, it grew fastest prior to 1970 with an annual average growth rate of 3.7%; its growth slowed down from 1990 to 2006 with an annual average growth rate of 1.7%. The order of annual average growth rate of different regions from top to bottom is as follows: Central and South America, North Africa, African countries south of the Sahara, Oceania and Asia (excluding China), all of whose annual average growth rates stood above the global average level. The annual average growth rates of the other regions were lower than the global average, among which the 12 countries of the former Soviet Union had the lowest growth rate of 0.4%.

The annual cereal output growth rates of developing countries and low-income food-deficit countries (LIFDCs) are 2.7% and 2.8% respectively, showing a fast growth and a gradual rising tend, whereas the annual cereal output growth rate of developed countries is merely 1.4%, lower than developing countries and LIFDCs, and the global average of 2%. The annual cereal output of developed countries went up from 0.54 billion tons in 1961 to 0.93 billion tons in 1978, and then stood at around 0.9 billion tons until 2006 when 0.98 billion tons of cereals were produced, with small fluctuations.

Major grain producers in the world include China, the U.S., India, France, Russia, Canada, Indonesia, Brazil, Argentina, Australia, Bangladesh, etc. The world's cereal output has grown rapidly since the beginning of the 1960s, and the world's cereal output of 2005 was twice to five times that of 1961. China's cereal output outstripped that of the U.S. after the 1980s and became the largest cereal producing country in the world. India was the third largest cereal producer, whose cereal output accounted for around 10% of the world's cereal output. The cereal output of the above-mentioned ten countries aggregated over 60% of the world's cereal output (see Fig. 8.2).

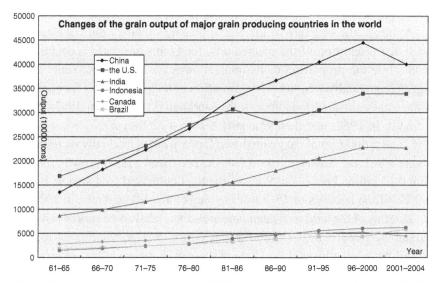

Figure 8.2. Changes of the Grain Output of Major Grain Producing Countries in the World. *Source: Data adapted in accordance with data of FAO.*

1.1.4. *The decreasing grain planted area due to agricultural industrial adjustment*

Through industrial structural adjustment, agriculture benefits could be improved. After solving the food problem, some countries started to adjust their agricultural industrial structures which have already become the main content in agricultural and rural economic development, and the cereal structures too changed continuously. In terms of cereal varieties, as the main plant food for countries throughout the world, rice, wheat and corn play an indispensable role in ensuring food security all over the world, and satisfying the global need for cereals, with their output accounting for around one-third of the aggregate cereal output respectively and their total output making up over 80% of the world's cereal output.

In terms of the aggregate cereal output, the proportion of the output of rice, wheat and corn in the aggregate cereal output was 25%, 25% and 23% respectively, the total of which is 73%; by 2006, the proportion of the output of rice, wheat and corn in the aggregate cereal output had reached 28%, 27% and 31% respectively. Hereinto, the proportion of corn output experienced the largest increase of 8%, with the sown area of corn going up by 43% from 0.1 billion hectares to 0.14 billion hectares and the corn output rising by 2 excess times from 0.21 billion tons to 0.64 billion tons. The proportions of rice output and wheat output did not witness much improvement.

Although the structure of the global planting industry did not change much, the proportion of the planted area for grains presented a declining trend while the proportion of the planted area for cash crops, fruits and vegetable gradually went up. The proportion of the planted area for cereals in the total cultivated land area in the world stood at 51% in 1962, rose to 52.9% in 1982, and then gradually fell to 47.2% in 2002, lower than 50%. The proportion of the planted area for cash crops gradually increased, among which the sown area of oil crops experienced a rapid growth, rising from 0.12 billion hectares or 9.1% of the total cultivated land area in 1961 to 0.22 billion hectares or 15.7% of the total cultivated land area in 2002.

In terms of the total cultivated land area, the global sown area of cereals had stabilized at around 0.7 billion hectares since 1961, accounting for about 50% of the total cultivated land area of the world; the sown area of cereals in developed countries had gradually decreased since the 1980s. The global sown area of cereals stood at 0.65 billion hectares in 1961, rose the peak at 0.73 billion hectares in 1981 and continued to fell after entering the 21st century to reach 0.67 billion hectares in 2006; it reduced by 0.05 billion tons in total during 1981–2006 with an annual average decrease of 2.4 million hectares, mainly because of the reduction in the sown area of cereals in developed countries due to food surplus. The global sown area of cereals in developed countries dropped by 0.08 billion hectares from 0.32 billion hectares to 0.23 billion hectares during 1981–2005, with an annual average decrease of 3.77 million hectares. Hereinto, the sown area of cereals in the U.S. reduced by 20.04 million hectares from 77.83 million hectares in 1981 to 57.79 hectares in the beginning of the 21st century with an annual average decrease of 0.9 million hectares; the sown area of cereals in the EU fell from 42.87 million hectares in 1981 to 26.86 million hectares in the beginning of the 21st century with an annual average decrease of 0.73 million hectares. Nevertheless, due to food shortage, the sown area of cereals in developing countries and the LIFDCs showed a rising trend.

1.1.5. *The rapid growth of world's per capita cereal output*

The per capita grain output represents the comprehensive agricultural strength of a region and is determined by the relative velocity between the changes in per capita grain output and the population increase. World's per capita cereal output showed an ascending trend prior to 1985 and then descended gradually. It stood at 310 kilograms in the 1960s and 363 kilograms in the 1980s, then declined gradually to 346 kilograms from 1991 onwards, and finally to 339 kilograms in 2006 due to the rapid population expansion, with an annual average increase of merely 0.3% since the 1990s.

The order of annual average growth rate of different regions from top to bottom is as follows: Central and South America, North Africa, Europe (excluding the twelve countries of the former Soviet Union), Oceania, the 12 countries of the former Soviet Union and North America, all of whose annual average growth rates stood above the global average level. The annual average growth rates of the other regions were all lower than the global average, among which China had the lowest growth rate of −0.1%. Hereinto, the order of the per capita cereal output of different regions from top to bottom is as follows: North America (1,442 kilograms), Central and South America (423 kilograms), Europe (excluding the 12 countries of the former

Soviet Union), Oceania and China, all of whose annual average growth rates stood above the global average level. The annual average growth rates of the other regions were all lower than the global average, among which African countries south of the Sahara had the lowest per capita cereal output of 84 kilograms.

The per capita cereal output of developing countries is merely 40% of developed countries. Because of sound agricultural foundation, rapid growth and small population, developed countries have a higher per capita cereal output. The per capita cereal output of developed countries stood at 656.2 kilograms in 1991 when the global average was 346.5 kilograms; whereas, the per capita cereal output of developing countries remained at 257.4 kilograms, 100 kilograms lower than the global average and accounting for merely 40% of that of developed countries. Meanwhile, the gap of the per capita cereal output between developed countries and developing countries is gradually widening, standing at 334 kilograms in the 1960s and expanding to 399 kilograms since 1991.

Great differences exist in the per capita cereal output among different countries. Canada, Australia, the U.S., France, Denmark and Hungary have high per capita cereal output of over 1000 kilograms. Since 1991, the per capita cereal output of the first four countries mentioned above was 1,664.7 kilograms, 1,573.2 kilograms, 1,170.4 kilograms and 1,041.0 kilograms, respectively. The per capita cereal output of Argentina, Germany, Ukraine and Poland stood between 500–1000 kilograms; that of Thailand, Burma and Britain equaled the global average; China, Brazil, India and Mexico stood below the global average; while Japan, Netherlands and South Korea maintained a low level.

1.1.6. *Grain output increase promoted by technological progresses*

Technological progresses are the main contributing factors of agricultural development and grain output increase, and the decisive factors in certain countries. A series of advanced agricultural technologies, such as, the Green Revolution, seed quality improvement, mechanization, chemistry, water conservancy, electric, information and ecology, and the establishment of the technological system significantly improved the overall grain productivity and guaranteed food security.

Since 1961, the aggregate cereal output went up by 1.4 times and the total sown area of cereals did not change much, increasing by 0.02 billion hectares or 3% from 0.65 billion hectares in 1961 to 0.67 billion hectares in the beginning of the 20th century; the per capita cereal output rose by 1.3 times from 1,353 kilograms per hectare in 1961 to 3,078 kilograms per hectare in the beginning of the 20th century, with an annual average growth rate of 2.0%. Therefore, the increase in the sown area for cereals made limited contributions to the increase of the aggregate cereal output, whereas the increase of per capita cereal output was the main contributing factor to the increase of the aggregate cereal output.

Scientific advancement and material input are the driving forces for increase in grain output. During the rapid growth period from 1961 to 1990, the global aggregate grain output increased by 1.3 times from 0.88 billion tons to 1.95 billion tons, and the per capita grain output increased by 1.1 times from 1353 kilograms per hectare to 2,775 kilograms per hectare. In terms of the increase of contributing factors, the sown area of cereals rose by 0.1 times; fertilizer consumption increased by 3.5 times with an annual average growth rate of 5.4%; the number of tractors ascended by 1.4 times with an annual average growth rate of 3.0%; the irrigation area rose by 0.8 times with an annual average growth rate of 2.0%. This indicates that the rapid growth of agricultural material input was the main driving force in the increase of the aggregate grain output and the per capita grain output while the slight increase of sown area contributed little to the increase of the aggregate grain output. Since the 1990s, the growth of aggregate cereal output had slowed down, rising from 1.95 billion tons to 2.03 billion tons with an annual average growth rate of 0.7%; the per capita cereal output ascended from 2,684 kilograms per hectare to 3,064 kilograms per hectare with an annual average growth rate of 1.2%; fertilizer input did not change much; agricultural machinery did not increase; the growth rate of irrigation area was 1.0%. Therefore, it can be concluded that the decline of material input growth rate was the main factor for the lingering aggregate cereal output.

Besides, the cereal output of developed countries lingered without pressing forward after 1986 with reduced fertilizer input and agricultural machinery input and unchanged irrigation area; the main reason for the decline of the growth rate of grain output was the reduced agricultural input due to the dampened enthusiasm for production; the cereal output of developing countries maintained a certain growth rate after 1990 mainly because the material input maintained a high growth rate.

1.1.7. *The potential and trends in world's food supply*

According to relevant research and predications prior to 2010, the annual global grain output would maintain a growth rate of 1.8% post-1990, and reach 2.34 billion tons in 2010; the growth rate of grain consumption during the same period would not exceed 1.6% and world's grain consumption would be expected to reach 2.23 billion tons in 2010. With the continuous increase of grain output, world's grain trading volume would also ascend continuously. It was also predicted that if the dependence degree of world's grain trade remained at the present levels, world's grain export volume would reach 0.33 billion tons and world's grain import volume would reach 0.32 billion tons in 2010, thus showing a trend of grain supply outstripping grain demand. FAO also believed that no unconquerable resource and technological restrictions exist in the world. Therefore, the growth rate of world's food output might exceed 1.8% in 2010, whereas the growth rate of world's food consumption will gradually descend to 1.3% in 2010 and to lower levels afterwards. In general, the growth in world's food production will outstrip the increase in food consumption worldwide.

Some people worry that the increase in food imports from the international market by China will trigger food crisis worldwide. This idea actually separates food supply from food demand. At present, there are abundant potential cultivated land resources in the world, including vast areas of unexploilted land in North America, Latin America, Australia, etc. According to U.N., there are about 0.36 billion hectares of idle arable land in the world, parts of which are easy to exploit, such as, the large amounts of fallow cultivated land in the U.S., Canada and the EU, the restoration of which will increase food output by around 40 million tons (Shi Peijun, 1999). If we use the major grain producing regions south of the Yangtze River as reference, the utilization rate and productivity of land in other countries of the world are all on the low side because these countries are unwilling to utilize their agricultural resources i.e. land intensively and the comparative advantage is on the low side. Once the so-called food crisis occurs, the rising food prices globally will rapidly mobilize the usable agricultural resources throughout the world (Lu Feng, 1999). In the past several decades, the global food market has been the buyers' market, with food prices globally showing a declining trend. If the demand for food goes up in the international market, the supply of food by grain producing and importing countries will increase correspondingly, thus substantially improving the gross supply of food in the world.

Besides, the scientific input and the application of biotechnology play an indispensable role in improving the potential of the per unit area agricultural yield. At present, the global average per unit area yield is not high. If the global average per unit area yield can reach the level of the U.S., the global aggregate grain output will reach 3.3 billion tons (Xu Gengsheng, 1998), which is sufficient to meet the global food demands. Once the improvement of the per unit area yield becomes profitable, corresponding technologies would soon emerge and be widely applied, thus promoting substantial increases in per unit area yield and the aggregate grain output. The exploitation and utilization of modern biotechnology provides a vast prospect for the rapid growth of the supply of food and other agricultural products.

1.2. Global food demand

The rapid development of agriculture promoted the noticeable improvement in the living standards of majority of residents and the continuous improvement of food security status in the world. The global per capita cereal consumption rose from 262 kilograms in 1961 to 314 kilograms in 2006. Serious food security problems concentrate in the underdeveloped regions of developing countries, such as, some regions in Africa and Asia. After entering the 1980s, food consumption in developed countries basically remained rather stable, whereas food demand in developing countries underwent a rapid growth, with a large gap between food production and food demand.

1.2.1. The continuous increase of gross cereal consumption

Since the 1990s, the global cereal consumption continued to rise, whereas the growth rate the global cereal output continued to decline, with the cereal demand outstripping the cereal output. In the four years after 2000, the global cereal consumption exceeded the global cereal output, causing the rapid decrease in the cereal stocks globally. Hereinto, the cereal demand of developed countries initially witnessed a rising trend and then a declining trend; the cereal consumption of developing countries firstly rocketed due to the rapid population expansion and income increase, and then showed a tendency toward stabilization after 1996. Ever since the worldwide food crisis occurred in the early 1970s, countries throughout the world paid much attention to food reserves which reached a record high of 0.46 billion tons in the year 1986–1987; after entering the late 1990s, with the decline of food output globally and the rise in food consumption, food stocks were used to balance the insufficient food production; the 0.32 billion tons of food stock globally in 2006 was 0.14 billion tons lower than and merely 70% of the highest record of accumulated food stock.

The aggregate annual global food consumption ascended from 1.47 billion tons in 1990 to 1.99 billion tons in 2006, with an average annual growth rate of 1.9%. The order of annual average growth rate of different regions from top to bottom is as follows: Central and South America, Oceania, North Africa, African countries south of the Sahara and North America, all of whose annual average growth rates stood above the global average level. The annual average growth rates of the other regions were all lower than the global average, among which the twelve countries of the former Soviet Union had the lowest growth rate of −1.7%.

The self-sufficient rates of different countries in the world presented a declining trend. According to the research by Wang Hongguan on the self-sufficient rates of over 140 countries in 2002 and 1982, the self-sufficient rates of 90 countries or 64.3% of the investigated 140 countries had declined since 1982, most of which were developed countries, including the U.S., Australia, Canada and Britain; the self-sufficient rates of 45 countries or 32.1% of the investigated countries had also declined since 1982, among which merely 24 countries or 17.1% of the investigated countries had self-sufficient rates of over 100% in 2002; the self-sufficient rates of 98 countries or 70% of the investigated countries stood below 95%; the self-sufficient rates of 73 countries or 52.1% of the investigated countries stood below 70%. China's self-sufficient rate has experienced a significant improvement, rising by 8.1% from 92.9% in 1982 to 101.0% in 2002.

Because of the decreasing cultivated land resources and the restraint in economic development, the self-sufficiency rates for cereals in some traditional grain importing countries further descended, and their dependence on the international food market was further enhanced.

Lingering between 260 to 340 kilograms, the annual average per capita cereal consumption of the world experienced an ascending trend and then a descending trend, rising from 262 kilograms in the early 1960s to 335 kilograms in 1986 and then dropping to 204 kilograms in 2002. The per capita cereal consumption of developed countries far exceeded the global average, going up from 468 kilograms (78.6% higher than the global average) in 1961 to 643 kilograms (almost twice that of the global average) in 1986 and then declining to 573 kilograms in 2002. The per capita cereal consumption of developing countries and LIFDCs stood below the global average and never exceeded 250 kilograms.

1.2.2. *Changes in cereal usage and the decline in the proportion of direct use of cereals as food*

The global cereal consumption proportion did not change much. Since 1961, 50% of the cereals have been directly consumed by human beings, 35% has been used as feed grain, and the remaining part has been used as raw materials, seeds, etc. Nevertheless, certain changes have taken place in the way cereal is used in developed countries, and the proportion of cereals directly used for food consumption has witnessed a gradual decline, dropping by nearly 10% from 31.6% in the 1960s to 22.8% in the early 21st century. For example, merely 13.2% of cereals were directly used for food consumption In the U.S.; this proportion fell to 12.2% in Canada and even lower in the Europe. The proportion of cereals used for feed grain experienced a slight increase in developed countries. For instance, 73.2% of cereals were used for feed consumption in Canada. No significant changes were reported in the proportion of cereals directly used for food consumption, which was over three times that of the developed countries. For example, over 90% of cereals were directly used for food consumption in Bangladesh and Ethiopia. Due to the development of animal husbandry, the proportion of cereals used for feed consumption also went up a little, rising from 8.5% in 1961 to 20% in the early 21st century.

1.2.3. *The improved food structure and slow development in developing countries*

Noticeable improvements have been made to the food structure in developing countries, which is, however, still rather simple, with a high proportion of starchy foods in the composition of energy and low nutritious quality. Researches on the scientific food structure suggest that starchy foods should provide about 55%–75% of the total calories provided by food consumption; if this proportion exceeds 70%, human bodies will suffer from malnutrition of trace elements and protein. Since the 1960s, the proportion of starchy foods in the food structure has been above 70% in over 20 developing countries, most of which are located in Africa and Southeast Asia; this proportion exceeds 76% in about 10 countries and 81% in three countries.

1.2.4. *Continuous improvement in the global food security status*

Since the 1960s, the proportion of the malnourished population of world has gradually descended by 20% from 37% to 17%, and the malnourished population of world has dropped from 0.96 billion to 0.8 billion. The proportion of the malnourished population of Asia experienced the largest decline of 24% from 40% to 16%. Hereinto, the proportion of the malnourished population of Southeast Asia and East Asia dropped by 33% from 43% between 1961 and 1971 to 10% in the early 21st century; the malnourished population of South Asia totaled 0.3 billion excess and accounted for 24% of the total population of this region; India had the largest malnourished population of 0.23 billion excess. Comparatively speaking, the proportion of malnourished population of Africa witnessed a slower decline of 6% from 34% to 28% while its population expanded from 0.11 billion to 0.2 billion. Hereinto, African countries south of the Sahara experienced the slowest decline and have a total malnourished population of 0.2 billion at present, being the most severely threatened region by starvation in the world. Economic poverty is the main reason for malnutrition. In general, due to deficient income and lack of food, food security problem is most outstanding in LIFDCs which have high proportions of malnourished population.

It is clear that the malnourished population in the world is concentrated in Asia, Africa, etc. Hereinto, Asia has over 60% of the malnourished population, and Africa has over 20%. The most severely threatened countries and regions by starvation in the world are mainly distributed in Africa and Asia, with nearly 60 million people facing food emergency situations. Food emergency status is more serious in some of the countries with natural disasters, civil wars and conflicts.

1.2.5. *The significant gap between developing countries and developed countries*

The aggregate output of agricultural products in the world can basically meet the need of human beings for food. However, the imbalanced distribution of resources, technology and wealth results in the deepening gap in food availability. Food surplus exists in some countries while food shortage and crisis occur in some developing countries. The developing countries face starvation with hundreds of millions of people suffering from hunger, whereas the developed countries face flat sales of food, bankruptcy of farms and the withering agriculture due to food production surplus.

Meanwhile, the gap between developed countries and developing countries is widening. The per capita supply of calories of developed countries is three times that of developing countries. The per capita cereal output gap between developed countries and developing countries shows an increasing trend, with the gap being nearly 400 kilograms. The per unit area yield gap between developed countries and developing countries continues to expand, with the gap exceeding 800 kilograms per hectare. The gap in scientific progresses between developed countries and developing countries also becomes rather significant.

2. World Food Trade

The global trading volume of food stood at merely 0.03 billion tons in 1946, doubled in 1956, reached almost 0.2 billion tons in 1979, and increased by three times during 1961–1981. The rapidly expanding world food trade plays a vital role in ensuring food security worldwide. Agricultural products trade with real objects as the main body is making up a decreasing proportion of the commodity trade in the world, whereas the value and volume of the global cereal trade goes up year by year, with a net growth of 2.4 times since the 1960s and an annual average growth rate of 3.2% which is higher than the annual growth rate of the global cereal production. Cereal exports concentrate in a few developed countries while cereal imports are rather dispersed. The cereal exports of the U.S. and the EU accounts for around 50% of the aggregate cereal exports of the world while Asia becomes the largest importer of cereals in the world whose cereal import makes up 40% of the aggregate cereal imports of the world. Since the 1980s, the import volume of grains by developing countries has witnessed a substantial increase, accounting for over 60% of the aggregate import volume of grain of the world.

2.1. *Changes of world cereal trading volume*

With the continuous increase in world's cereal output, corresponding changes happen to world cereal import and export (see Fig. 8.3). The value and volume of world's cereal trade went up year-by-year, with world cereal trading volume ascending by seven times and reaching nearly $100 billion since the 1960s, and the annual average growth rate standing at 5.1%. Hereinto, the growth rate of world's cereal import (6.4%) was slightly higher than that of world's cereal export (6.4%); the trading volume of cereals experienced a net increase of 2.4 times with an annual average growth rate of 3.0%.

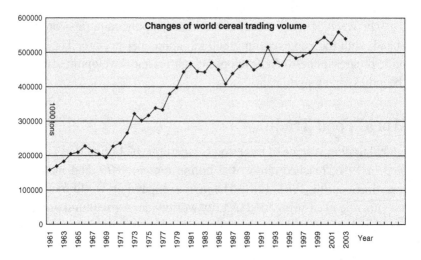

Figure 8.3. Changes in World's Cereal Trading Volume.
Source: Data adapted in accordance with data of FAO.

2.1.1. *World cereal circulation amounting to nearly 0.3 billion tons*

Since the 1960s, world cereal trading volume has gone up continuously, rising from 0.2 billion tons in the 1960s to 0.3 billion tons in the 1970s, 0.45 billion tons in the 1980s, 4.9 billion tons in the 1990s, over 0.5 billion tons in the beginning of the 21st century and 0.54 billion tons in recent years. The global cereal import generally equaled the global cereal export with a cereal circulation of 0.27 billion tons, accounting for 13% of the global cereal output. Among the trading volume of cereals, rice, wheat and corn account for the largest proportions. Wheat has been the largest trading grain variety, and its trading volume stood at nearly 0.11 billion tons and accounted for 41% of the aggregate trading volume of cereal of the world. Corn is the second largest trading grain variety, and with a trading volume of nearly 0.09 billion tons it constituted 32% of the aggregate trading volume of cereal of the world. Rice is the third largest trading grain variety, the trading volume of which stood at nearly 0.03 billion tons and took 10% of the aggregate trading volume of cereal of the world. In terms of importing and exporting countries, cereals are mostly exported by a few developed countries, with the export volume of cereals by the U.S. accounting for 30% of the aggregate export volume of the world and the export volume of cereals by the EU accounting for 20% of the aggregate export volume of the world; cereals are mostly exported by Asian countries, with the import volume of cereals of these countries making up 40% of the aggregate import volume of cereals of the world; hereinto, the import volume of cereals of Japan, China and Korean accounted for 20% of the aggregate import volume of cereals of the world.

2.1.2. *Developing countries as the main importers of cereals*

The world's cereal import maintained a momentum of growth, rising by 2.4 times from 79.796 million tons in 1961 to the most recent 268.68 million tons. After entering the 1990s, the import volume of cereals by developed countries remained rather stable, standing around 90 to 100 million tons; the demand for cereals by developing countries rapidly expanded, resulting in both the increase of domestic cereals production and the rapid growth of import volume cereals which reached 0.17 billion tons in the beginning of

Food Security and Farm Land Protection in China

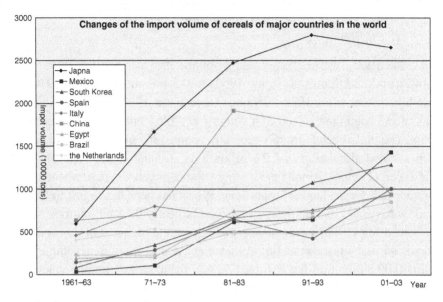

Figure 8.4. Changes of the Import Volume of Cereals of Major Countries in the World.
Source: Data adapted in accordance with data of FAO.

the 21st century, accounting for over 60% of the aggregate import volume of grain of the world. The growth rate of the import volume of cereals, however, showed a declining trend. In the 1970s, the import and export of cereals of the world underwent a fast growth with an annual average growth rate of 7%; hereinto, the annual average growth rate of the import volume of cereals by developed countries reached 5.8% whereas that of developing countries stood at 8.8%. In the 1990s, the import volume of cereals of the world grew slowly with an annual average growth rate of 1.7%; hereinto, developed countries had a negative annual average growth rate of −0.3%; whereas the annual average growth rate of developing countries stood at 3.2%. Asia has the fasted growth of cereal imports, with its import volume of cereals in the beginning of the 21st century being 2.5 times that of the 1970s and accounting for 40% of the aggregate import volume of cereals of the world. The import volume of cereals of the EU made up 22% of the aggregate import volume of cereals of the world and maintained a low growth rate.

The major importers of cereals in the world include Japan, Mexico, South Korea, Span, China, Egypt, Italy, Brazil and the Netherlands, the total import volume of cereals of which made up 40% of the aggregate import volume of cereals of the world (see Fig. 8.4).

2.1.3. *A few developed countries as the major exporters of cereals*

With the increase in output of cereals globally, the export volume of cereals has increased year-by-year. Developed countries are the major exporting countries of cereals. Since the 1960s, the export volume of cereals globally had risen by 2.4 times with an annual average growth rate of 3.0%. Developed countries accounted for a large proportion of over 80% of the export volume of cereals of the world prior to 1996; this proportion then fell slightly to 70%. The export volume of cereals of a few developed countries made up around 60% of the export volume of cereals globally, including the U.S., which is the the largest exporter of cereals and whose export volume of cereals accounted for nearly 30% of the global export volumes (of cereals). Similarly, France's export volume of cereals accounted for nearly 10% of the export volume of cereals of the world. In comparison with developed countries, the export volume of cereals of developing countries made up a small proportion of the export volume of cereals of the world. Since 1961, the export volume of cereals of developing countries went up by over five times, and the growth rate also witnessed a slight increase. China and India had large export volumes of cereals, which accounted for 5% and over 3% of the global export volumes (of cereals) (see Fig. 8.5).

2.1.4. *The insignificant gap between world's food supply and demand*

In 2006, world's food output dropped slightly and the gap between food production and demand in the market continued to increase, leading to a significant rise in international food prices. According to data of the recent 20 years, the world's food consumption slightly outstripped the world's food output in most years, with the highest ratio of the world food consumption to the world's food output being 105% and the average annual ratio being 102.2%. This ratio stood below one in one-third of these years.

Adverse weather was the major reason for the reduction in the global food output. The global food consumption experienced a steady growth,

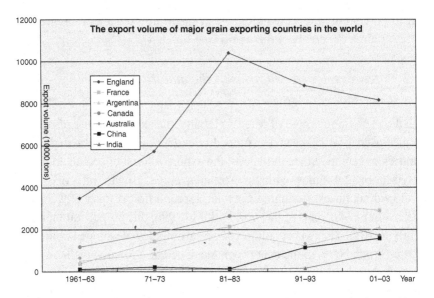

Figure 8.5. The Export Volume of Major Grain Exporting Countries in the World. *Source: Data adapted in accordance with data of FAO.*

with the corn demand outstripping the corn output and the consumption of corn and rice witnessing a rapid increase. In addition, developed countries encountered food output reduction successively, leading to the substantial decrease in world's food reserve which sometimes dropped to below the minimum safe level of 18%. Because of the reduction in food output and the increase in food demand, world's food reserve started to decline and the food supply-demand relationship in the international market was tightened, leading to the substantial rise in food prices and in the prices of food futures in the international market.

It can be seen that the food demand continued to outstrip the food output in the international market and the food stock decreased substantially, and that the international food market picked up significantly due to the increasing pressure on the market brought by the gap between food demand and output. This consequence is expected to further enlarge the gap between food demand and output globally. Therefore, food output must be increased so as to reduce, if not completely eradicate, the historically accumulated gap.

2.2. *Changes of cereal varieties of international trade*

Wheat, corn and rice are three major cereal varieties in international trade which have been occupying important positions in cereal trade and grain trade.

2.2.1. *The large proportions of wheat, rice and corn in the trading volume of cereals*

Wheat, corn and rice are three major varieties of international trade of cereal. Wheat has been occupying the largest proportion in the trading volume of cereals, and corn and rice have been occupying the second and third largest proportions, the total of which accounts for over 75% of the trading volume of cereals. Since the 1960s, the trading volumes of wheat, corn and rice have shown a rising trend, with the export volume of the three rising from 75.3% in 1961 to the most recent 83.1%. As the cereal variety with the largest trading volume, the export volume of wheat reached 0.11 billion tons in the beginning of the 20th century; the trading volume of corn reached 0.09 billion tons, accounting for 32% of the trading volume of cereals; the trading volume of rice was comparatively lower than that of wheat and corn, standing at 0.03 billion tons and constituting around 10% of the trading volume of cereals in the world (see Fig. 8.6).

2.2.2. *Export volume of wheat accounting for over 40% of the total export volume of cereals*

Wheat occupied the largest proportion of the trading volume of cereals, with about one-fifth of the world's wheat output being used for export. Affected by natural conditions, the export volume of wheat had large fluctuations. Since the 1960s, the proportion of the trading volume of wheat (in the overall trading volume of cereals) has declined continuously. The proportion of the export volume of wheat in the total export volume of cereals has dropped from 50% to 41%; despite of the continuous decline of the proportion of the trading volume of wheat in the overall trading volume of cereals, the trading volume of wheat has increased continuously, with the export volume of wheat reaching 0.11 billion tons in 2006, 1.8 times that of 1961.

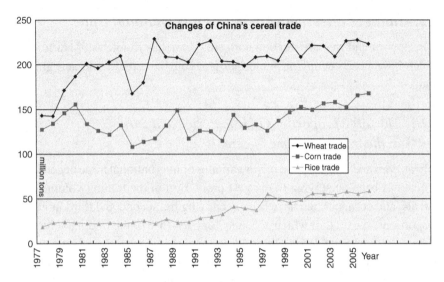

Figure 8.6. Changes in China's Cereal Trade.
Source: *Data adapted in accordance with data of FAO.*

In terms of regions, Asia is the largest importing region of wheat. After entering the 21st century, the annual import volume of wheat by Asia reached nearly 0.04 billion tons, accounting for over 35% of the total import volume of wheat of the world; Europe also imported large amounts of wheat, with its import volume of wheat making up over 20% of the total import volume of wheat of the world; the import volume of wheat by Africa ranked third, standing at 23 million excess tons. In terms of countries, Italy was the largest importing country of wheat with an annual average wheat import of 7 million excess tons, accounting for 6.5% of the total import volume of wheat of the world; Brazil was the second largest importing country of wheat with an annual average wheat import of 6.7 million excess tons; Japan ranked third with an annual average wheat import of 5.5 million excess tons, making up 4.9% of the total import volume of wheat of the world. Europe was the largest exporting region of wheat with the largest export volume of wheat; its annual average export volume of wheat has reached 42.22 million tons since the beginning of the 21st century, amounting to 37% of the total export volume of wheat globally. North America followed close behind with an annual average wheat export of 39 million excess tons. In terms of countries, the U.S. was the largest exporting country of wheat with an annual average export volume of 25.15 million tons, accounting for over 21% of the total export volume of wheat globally; Canada ranked second with an annual average wheat export of 15 million excess tons, making up over 12% of the total export volume of wheat of the world.

2.2.3. *The rapid growth in the trading volume of corn*

The trading volume of corn took the second largest proportion of the total trading volume of cereals. After World War II, the development of global animal husbandry promoted the expansion of corn trade globally, and corn had a larger market as one of the major feed sources. Since 1961, the proportion of the trading volume of corn in the total trading volume of cereals has shown a rising trend year by year, with the proportion of the export volume of corn in the total export volume of cereals going up from 17.6% to 32.3% and the export volume of corn increasing continuously to reach 0.08 billion tons in 2006, 4.9 times that of 1961.

Since the beginning of the 21st century, Asia has become the largest importing region of corn in the world with an annual average corn import of 0.05 billion excess tons, accounting for 50% of the total import volume of corn of the world; the import volume of corn of Latin America and the Caribbean Sea area made up 17% of the total import volume of corn of the world with an annual average corn import of nearly 15 million tons. In terms of countries, Japan was one of the largest importers of corn in the world with an annual average corn import of 16 million excess tons, accounting for nearly 20% of the total trading volume of corn of the world; South Korea ranked second with an annual average corn import of 8.8 million tons, accounting for over 10% of the total trading volume of corn of the world.

North America is the major exporting region of corn while the other continents are net importing regions of corn. North America exported an average of 46.5 million tons of corn annually since the beginning of 21st century, accounting for over 55% of the total export volume of corn of the world; Asia ranked second with an annual average corn export of 12 million excess tons, accounting for 14.6% of the total global export volume of corn. The U.S. was the largest corn producing country with an annual average corn export of 46.35 million tons, accounting for around 55% of the total export volume of corn globally; China ranked second with an annual average corn export of 11.36 tons, accounting for 13.6% of the total export volume of corn globally.

2.2.4. *The trading volume of rice occupying a certain position*

The trading volume of rice fell behind that of wheat and corn and ranked third. Since 1961, the proportion of the trading volume of rice in the total global trading volume of cereals has shown an ascending trend year by year, with the proportion of the export volume of rice in the total export volume of cereals globally increasing from 7.9% to 10.2% and the export volume of rice increasing by 3.1 times over that of 1961 to reach 25.5 million tons in 2006.

As one of the major producing regions of rice in the world, the rice output of Asia accounted for over 90% of the total rice output in the world, and its trading volume of rice occupied a high proportion of the total trading volume of rice in the world. Asian region was also the largest importing region of rice in the world. Since the beginning of the 21st century, Asia imported an average of 11.41 million tons of rice annually, accounting for 56.2% of the total import volume of rice in the world; Africa was the second largest importing region of rice with an annual average rice import of 6.5 million excess tons, accounting for 32.3% of the total import volume of rice of the world. In terms of countries, India was the largest importer of rice in the world with an annual average rice import of 1.36 million tons, making up 5.4% of the total import volume of rice globally.

Asia was also one of the major exporting regions of rice in the world. From 2001 to 2006, the annual average rice export of Asia reached 20 million excess tons, accounting for 74.2% of the total export volume of rice globally; North America ranked second with an annual average rice export of 3.23 million tons, making up 11.8% of the total export volume of rice globally. In terms of countries, Thailand was the largest exporting country of rice with an annual average rice export of 7.81 million tons, accounting for 28.6% of the total export volume of rice globally; Vietnam was the second largest exporting country of rice with an annual average rice export of 3.59 million tons, accounting for 13.2% of the total export volume of rice globally.

2.3. *The correlation between world's cereal output and world's cereal trading volume*

The correlation coefficient between world's cereal output and world's cereal trading volume was 0.5 between 1961 and 2006. Generally speaking, the average correlation between world's cereal output and world's cereal trading volume indicated that the cereal trading volume increases along with the growth of cereal output, but it failed to make a sound explanation of their relationship. In terms of different periods, the correlation coefficient was 1.0 during 1961–1970, 0.9 during 1971–1980, 0.8 during 1981–1990, 0.7 during 1991–2000 and 0.8 during 2001–2006. In terms of the correlation coefficients of different periods, the correlation coefficient stood above 0.9 both in the 1960s and the 1970s, fell slightly in the 1980s and the 1990s and did not pick up even after 2000. It is clear that the global trading volumes in food maintained a sound correlation with world's food output in the 1960s and the 1970s and that the correlation coefficient started to fell after entering the 1980s due to the increase in trade barriers. The correlation coefficients of the period prior to 1980, the period from 1980 to 2000 and the period after 1981 was 1.0, 0.8 and 0.6 respectively, indicating that world's trading volume in food was significantly affected by world's food output prior to 1980 and the correlation coefficient declined in the 1980s and 1990s and that the correlation between world's food trading volume and world's food output was even weaker after 1981.

The ratio of world's cereal import volume to world's food output corresponded with the above research results. Figure 8.7 shows that the ratio of world's cereal import volume to world's food output basically presented a declining trend, falling from 10% in the mid-1970s to 8% in the 1980s, 7% in the 1990s and 6% in the beginning of the 21st century.

2.4. *The influence of the production of ethanol from corn in the U.S. on world's food market*

Food prices across the globe have soared since 2007 and hit a new high in 20 years. Wheat price jumped by 112%, and corn price rose by 47.3%. Soybean price and rice price followed, accompanied by food panic and disturbances in many countries. These events were directly related to the promotion of production of ethanol from corn in the U.S. As one-fourth of the corn output

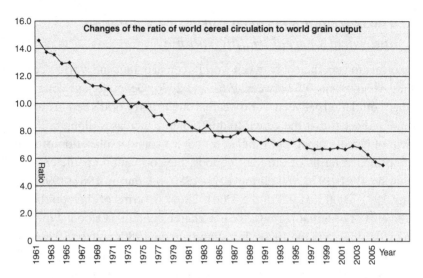

Figure 8.7.　Changes in the Ratios of World Cereal Circulation to World Grain Output.
Source: Data adapted in accordance with data of FAO.

of the U.S. was used to produce ethanol, it was expected that the corn export of the U.S. would decrease by 48% in 2008. The corn export of the U.S. made up 75% of world corn export. Meanwhile, since corn price was increasing rapidly, large numbers of American farmers started to plant corn instead of soybean, with the sown area of soybean in the U.S. decreasing substantially by 15.6%. The planted area for corn in the U.S. reached 93.6 million acres (1 acre equals to 0.4 hectare) in 2007, a new high since 1944, and the corn output of that year stood at 12.5 billion bushels. Although the corn output hit a new high, it was expected that the corn price would continue to rise to a record high of &3.5 to &3.6 per bushel. The U.S. produced 4.2 billion gallons (1 gallon equals to about 3.8 liter) of ethanol biofuel in 2005 and 8 billion gallons of ethanol biofuel in 2007 which could replace 2% of its oil use. Based on current production technology, 1 bushel of corn could generate 2.7 gallons of ethanol. In this way, 2.96 billion bushels of corn was needed for the production of 8 billion gallons of ethanol. As the annual corn output of the U.S. stands around 12 billion bushels, it has a large potential.

As the production costs of corn-based ethanol could be covered when oil price reaches $50 per barrel and the current oil price stands above $100 per barrel, the production of ethanol from corn will make considerable profits. The oil price of $100 per barrel or 42 gallons equals to $2.4 per gallon. According to the estimation of Keith Collins, the chief economist at the Department of Agriculture of the U.S., as the production cost of corn-based ethanol is 1.6 dollars per gallon, the selling price of corn-based ethanol is 51 cents higher than that of gasoline and the federal governments reduced the tax on blend fuel producing companies by 51 cents per gallon, thereby enabling the total rate of profit of corn-based ethanol to reach nearly 50%. The production of corn-based ethanol is profitable even after the intermediate links are taken into consideration. According to a report released in 2007 by the Global Subsidies Initiative (GSI) Geneva, the subsidy for biofuel given by the U.S. government stood between $6.3 billion to $7.7 billion in 2006 and is expected to reach $13 billion in 2008 and $16 billion in 2014. It was reported that the number of ethanol factories in the U.S. was 59 in 2001 and rose to 119 in 2007 with another 86 factories being built, among which are certain well-known global enterprises. 30% of gasoline in the U.S. will be replaced by liquid biofuels, the total amount of which will reach 60 billion gallons. If the U.S. demand for oil grows by 1% to 2% annually, this can be met by the 60 billion gallons of liquid biofuels, from now until 2030, and the U.S. will be able to guarantee its oil security as long as it maintains the present oil import volume.

3. WTO and Agricultural Trade

3.1. *China's commitments in the WTO transitional period*

According to the classification of WTO, food trade policies include domestic support, market access and export competition. Since its entry into the WTO, China's food trade polities mainly include the following aspects:

3.1.1. *Policies on domestic support*

China promised a *de minimis* provision[1] of 8.5% for domestic support for agricultural products when she entered the WTO. The current support level

---

[1]The *de minimis* provisions: there is no requirement to reduce trade-distorting domestic support in which the aggregate value of the support does not exceed 5% of the total value of production of the agricultural product in question (in the case of developing countries the *de minimis* ceiling is 10%).

in China is merely 0.6%, far lower than the promised scope. Though China has made certain adjustments in its policies relevant to domestic support, the following aspects are noticeable.

The first aspect is "two tax reductions and three subsidies". Two tax reductions referred to the exemption of agricultural tax and tax on special agricultural products excluding tobacco leaves. Three subsidies referred to direct food subsidies, seed subsidies and agricultural machinery subsidies. China abolished agricultural tax in an all-round way in 2006, two years ahead of its five-year schedule. Three subsidies were increased from 0.1 billion RMB in 2002 to 30 million excess RMB in 2006.

The second aspect is the implementation of grain minimum purchase price policy in major grain-producing regions. In order to protect peasants' enthusiasm for grain production in major grain-producing regions, grain minimum purchase price policy was continued in these regions, with paddy rice and wheat as the main targets. Although violating the WTO rules, the minimum purchase price was far lower than China's *de minimis* provision. This policy provided peasants with a signal of stable grain price expectation, fully aroused their enthusiasm for grain production, and protected their interests.

The third aspect is the integration of agricultural insurance in the system of support and protection for agriculture. At the end of 2006, China decided to integrate agricultural insurance to support and protect agriculture and openly put forward the "three subsidies" policy which subsidized rural households, insurance enterprises and agricultural reinsurance. The main contents of the "three subsidies" policy included: the central and local finance should offer proportional subsidies to rural households in accordance with the agricultural insurance varieties; the central and local finance should offer appropriate subsidies for the operating expense of policy-oriented agricultural insurance to insurance enterprises; the system of agricultural reinsurance supported by the central and local finance should be established. Through its direct support to the development of agricultural insurance, the government indirectly offered policy support and protections of interests to local agriculture and local households. Surely, agricultural insurance belongs to one of "the green box policies" permitted by WTO and is one of the important means of supporting and protecting agricultural development used by developed countries.

3.1.2. *Policies on market access*

China implemented tariff quota management on staple agricultural products, such as, grains (wheat, corn and rice), cotton, edible vegetable oils, sugar and wool, and tariff-only management on soybean, horticultural products, animal products, and other agricultural products. In 2004, the weighted-average tariff rate for agricultural products was 8%, far lower than its commitment of lowering the tariff rate for agricultural products from 21.1% in 2002 to 15% in 2004 when entering the WTO. The import tariff for agricultural products was further reduced in 2005. In fact, China's present tariff quota policies have not exercise the function of limiting import. China had the largest food import of 80.2 million tons in 2004 since its entry into the WTO with a quota application rate of merely 36%.

3.1.3. *Policies on export competition*

At the time of joining the WTO, China had promised to abolish all export subsidies. Since the end of 2001, the Chinese government had taken two other major relevant measures: The abolishment of railway construction fund and export rebates. First, railway construction fund was abolished so as to adjust its influence on food market. China decided to fully exempt the transportation of paddy rice, wheat, rice, wheat flour, corn and soybean by railways from railway construction fund starting from April 1, 2002. It was also stipulated that the time limit for the implementation of the exemption measures was tentatively set at the end of 2005 and no similar policies regarding reduction of railway construction fund would be issued from then on. The exemption period was extended. As the railway construction fund made up 30%–40% of the total transportation costs, the exemption measures reduced the transportation cost from grain producing regions to grain selling regions (by rail) by 40% in average. Second, in terms of export rebates, the State Council approved that the export of rice, wheat and corn enjoyed a zero rate of VAT and was exempt from the output tax on April 1, 2002. The new export rebate policies issued in 2005 made a supplementary regulation that the export rebate rate for wheat flour and corn flour was raised from 5%–13%. Export rebates completely comply with the rules of the WTO and aim at strengthening the international competitiveness of China's exported products.

3.2. *China's commitments in post-WTO transitional period*

Starting from 2005, most of the transitional periods for major industries obtained in the negotiation for China's entry into the WTO have come to an end. After entering the post-WTO transitional period, China's foreign trade is experiencing new development trends. The basic situation of China's fulfillment of its commitments and enjoyment of its rights in the post-WTO transitional period is as follows. In accordance with its WTO commitments, China will further open up its market in the post-WTO transitional period.

First, the import tariff will be reduced to the ultimate promised level. The overall tariff level had been further lowered from 10.4% in 2004 to 9.9% in 2005, among which the average tariff for industrial products was lowered from 9.5% to 9% and the average tariff for agricultural products was reduced from 15.6% to 15.3%.

Second, all the non-tariff measures will be abolished. All non-tariff measures, including import quota, import license and specific products tendering, have been abolished. The import quota management on automobiles, the most sensitive industrial product in the negotiation for China's entry into the WTO was also abolished and replaced by automatic import license.

Third, the amount of tariff quotas for agricultural products excluding vegetable oil (including soybean oil, palm oil and rapeseed oil), has reached the highest level since 2004. The system of designated management of the import of wool and wool tops was abolished in 2005. Tariff quotas for vegetable oil was abolished and replaced by the regulation of import by tariffs on 1 January 2006.

Fourth, trade rights have been fully opened. Starting from the second half of 2004, all individuals and enterprises in China were permitted to engage in import and export business.

Fifth, limitations in most fields of service trade will be abolished and the establishment of foreign-controlled and wholly foreign-funded enterprises in the service trade was permitted. Hereinto, in the banking sector, China has removed all the regional limitations and client scope limitations on the RMB business of foreign banks since 11 December 2006; in the insurance sector, compulsory reinsurance has been abolished since 2005, and the establishment of wholly foreign-funded insurance brokers companies has been permitted since 2006; in the mobile voice and data service of telecommunication departments, a foreign capital share of 49% was permitted at the end of 2004 and the regional limitations were abolished at the end of 2006. It was announced that in both the domestic and international business of basic telecommunication, regional limitations will be abolished and a foreign capita share of 49% will be permitted by the end of 2007; in construction design, tourism and transportation departments, the establishment of wholly foreign-funded enterprises will be progressively allowed from 2005 to 2007.

Finally, China will accept deliberation of its trade policies according to the principle of transparency. Apart from fulfilling its obligations of informing, transitional deliberating, consulting and providing comments before laws and regulations affecting international trade are implemented, the WTO will conduct the first deliberation of China's trade policies in April 2006, after which China will be deliberated every two years as one of the four major trading countries of the WTO.

Apart from fulfilling the above commitments and the obligation of further opening its market, China will also further enjoy the rights of members of the WTO, including the following major aspects. First, China will enjoy the most-favored-nation treatment and national treatment accorded by other members of the WTO and benefit from the abolishment of worldwide textile quotas and the global textile trade integration from 2005. Second, through its participation in negotiations of the WTO, China will directly get involved in the formulation of multilateral economic and trade rules and fully play its role in international economic and trade affairs. Third, China will negotiate with new members applying to join in the WTO, on market access. Fourth, China will enjoy rights of the members of developing countries. Fifth, China will obtain the protection of multilateral trade system and solve its disputes with other member countries in trade and investment by making use of the dispute settlement mechanism of the WTO.

3.3. Several important agricultural trade agreements in the world

3.3.1. The WTO's agreement on agriculture

According to the Impact of the WTO's Agreement on Agriculture on China's Agricultural Development, the WTO's Agreement on Agriculture was implemented from January 1, 1995 to December 31, 2000 (10 years for developing countries). Members also agreed to carry out talks one year prior to the completion of the implementation period so as to continue the reforms. In summary, the agreement allowed its member-governments to grant subsidies to agriculture and encouraged the use of less trade-distorting domestic support policies. The agreement also allowed some flexibility in the implementation of commitments. The reduction of subsidies and tariffs by developing countries did not have to equal to that of developed countries, and developing countries were granted more time to fulfill their commitments. Specific concerns of developing countries have also been addressed including the concerns of net-food importing countries and least-developed countries.

✧ Market access provisions. The restriction of the entry of agricultural products into their domestic market through tariffs and various non-tariff barriers by many countries (especially developed countries) resulted in an unfair competition in the international trade of agricultural products and hindered the realization of liberalization of the international trade of agricultural products. In view of this, the WTO's Agriculture Agreement required that all its members should try their best to remove the interference of non-tariff measures and adopted the provisions of tariffication of non-tariff barriers and prohibition of new non-tariff barriers so as to reduce the existing non-tariff barriers in agricultural trade. In addition, the WTO members also reached agreements on increasing market access opportunities for agricultural products so as to promote the realization of liberalization of the international trade of agricultural products. The specific provisions are as follows:

• The tariffication provisions permitted restrictions on agricultural products trade only in the form of tariffs and requested that quantitative import restrictions, import variable duties, minimum import prices, import licenses, non-tariff measures maintained by units in charge of management of goods under state monopoly, voluntary export restrictions, similar border measures apart from ordinary tariffs and other non-tariff measures should all be transformed into import tariffs.

• Tariff concession provisions request that all WTO members should promise to reduce their tariffs of the base period (including the new tariff rates after tariffication) to certain levels during the implementation period.

- Minimum market access provisions stipulate that the WTO members should promise to ensure a minimum access opportunity when their imports of tariffied products during the 1986–1988 base period had been less than 5% of domestic consumption of the product in question (3% for developing countries). In the first year of the implementation period of the agreement, WTO members should ensure that current and minimum access opportunities combined represented at least 3% of base-period consumption and were progressively expanded to reach 5% of that consumption by the end of the implementation period. The minimum access opportunities are generally implemented in the form of tariff quotas. That is to say, in order to ensure the imports of agricultural products of the minimum access opportunities by the WTO members, all WTO members should ensure a low or minimal applicable duty on imports of the minimum access opportunities, but could apply higher customs duty after tariffication to any imports outside the tariff quota.

- The agreement requested that its member countries should maintain the existing market access of the base period when the current market access is against the minimum access requirement by the agreement, or when the imports of a certain agricultural product by its member countries exceeded 5 percent of the domestic consumption of this product in question (3 percent for developing countries). That is to say, its member countries should maintain the existing tariff rates.

- The special safeguard provisions build a special safeguard mechanism for agricultural products required to be tariffied. That is, the special safeguard provisions allow the imposition of an additional tariff when a specified surge in imports of agricultural products or a fall of the import price below a specified reference price occurs.

- Special and differentiated treatment relaxes the requirements on market access for developing countries.

✧ Domestic Support Provisions. It is important that members (countries/ regions) implement measures to support domestic agricultural production as lack of such measures give rise to the unfair competition in international trade of agricultural products. Negotiations on the international trade of agricultural products in the Uruguay Round included tough and detailed discussions on how to distinguish "trade-distorting production measures" and "nontrade-distorting production measures" and finally divided different domestic support measures into two categories. The first category includes measures with no distortive effect on trade which are often referred to as "Green" measures or "Green Box" measures and are exempt from reduction commitments. The second category includes trade-distorting measures which are often referred as "Amber Box" measures. The agreement requested that the aggregate monetary value of Amber Box measures should be subjected to reduction commitments as specified in the schedule of each WTO Member providing such support.

• The Green Box measures. The WTO's Agriculture Agreement stipulates that the general criteria of the Green Box should include measures that have no, or at most minimal, trade-distorting effects or effects on production; they must be provided through a publicly-funded government program (including government revenue foregone) not involving transfers from consumers and must not have the effect of providing price support to producers, and that subsidies belonging to these measures would be considered as green subsidies and would be exempt from reduction commitments.

• Amber Box Policies. Measures promised by all member countries regarding reduction and restriction of domestic agricultural support and subsidy in accordance with requirements of the WTO mainly refer to policy measures which easily distorts trade of agricultural products, including the governments' direct intervention in the price of and subsidies to agricultural products, seeds, fertilizers, irrigation and other agricultural inputs, subsidies to marketing of agricultural products and fallow land. Such policies are generally called "Amber Box Policies". Agricultural subsidies under the "Amber Box Policies" are called "Amber Box Policy" subsidies. The WTO's Agriculture Agreement stipulates that subsidies under the Amber Box should be calculated under the Aggregate Measurement of Support (AMS) and are subject to reduction discipline on the basis of constraints on this sort of subsidies.

- Export subsidy provisions. Export subsidy refers to subsidies contingent on export performance. It is the policy measure which most easily causes unfair competition (is the most trade-distorting). Negotiations prior to the Doha Round succeeded only in imposing limitations on export subsidies for industrial products. The Doha Round negotiations made certain progress in the reduction of agricultural export subsidies and agreed on relevant provisions on using the reduction of export subsidies for the base period as the standard and cutting agricultural export subsidies gradually within a certain implementation period.
- Sanitary and Phytosanitary (SPS) Measures provisions. Environmental protection and SPS measures in the international trade of agricultural products refer to certain measures taken by its member countries (regions) to restrict agricultural product imports for the sake of protecting human, animal or plant life safety and health. This sort of import restriction measures has certain rationality. However, the arbitrary use of this sort of measures to build trade barriers prevailed in the international trade of agricultural products in recent years. In view of this, the WTO's Agreement on Agriculture stipulates that the following: (1) Restriction in the import of agricultural products in disguised forms with environmental protection and protection of animal or plant life or health as excuses is prohibited; (2) SPS measures imposed on agricultural products must be based on science (international standards or norms); (3) Member countries could adopt interim SPS measures on the basis of relevant existing information when scientific evidence is weak; and (4) all these export limitation measures should be taken on the premise of high transparency.

3.3.2. *The doha round of agricultural trade negotiations*

The "Doha Round" of agricultural negotiations has three major aims: Substantially increased market access for agricultural products; elimination of export subsidies in various forms by phases; and substantial reductions in various domestic subsidies.

At the WTO's Fifth Ministerial Conference in September 2003 (or the WTO's Cancún Ministerial Conference), agricultural subsidies became the focus of discussion among all parties. Developing country members strongly urged the Europe and the U.S. to abolish trade-distorting agricultural subsidies and to provide a detailed timetable and requested the U.S. to revise its Agricultural Act unilaterally. The Europe and the U.S. requested developing members to reduce their tariffs and enlarge the market access for products of other countries. In terms of the discussion order of conference themes, developing country members insisted that the theme of agriculture should be discussed first before other themes could be discussed, whereas developed country members were anxious to discuss the initiation of "Singapore Issues" negotiations. The two sides came to a deadlock and the conference ended with no results.

In March 2004, "Doha Round" of negotiations on many themes was resumed. After four months of fierce quarrels, the WTO members finally approved a Framework Agreement on the major themes of the Doha development agenda in August 2004. Regarding the crucial issue of negotiations — agriculture, developed countries finally committed to abolish export subsidies, reduce domestic support and substantially improve market access conditions in the Framework Agreement on Agriculture. The framework of Doha Round agricultural negotiations formulated the road map and some key standards of the three pillars of agricultural negotiations — substantial improvements in market access, gradual reduction or elimination of all forms of export subsidies and substantial reductions in trade distorting support. However, no more substantial progress has been made in agricultural negotiations since then.

The WTO family was naturally divided into three interest groups, namely, the G-20 which represents developing countries, the Cairns Group of agricultural exporting countries represented by the U.S. and the group of agricultural self-sufficient countries represented by the EU, Japan and South Korea. In the agricultural negotiations, the three sides carried out a contest of strength for interests. In a large number of developing countries, 38% (71% in the least developed countries) of the labor force is employed in the agricultural sector, and the agricultural trade usually accounts for over 50% of the export volume of these countries. As the majority of the impoverished population of developing countries lives in rural areas and engaged in farming, agriculture is of crucial significance to the impoverished population. On the contrary, developed countries represented by the U.S. not only have advanced agricultural technology but also provide their farmers with an annual subsidy of $300 billion, thus completely depriving the agricultural products of developing countries of their price advantages in the international market. Data shows that developing countries will be able to obtain an annual economic welfare of $43 billion if all trade barriers in the agricultural and food sectors are removed. Producers and exporters in the developing countries will derive more benefits from the rising prices of agricultural products and fairer competitions. Although with strong competitiveness in agricultural products, the U.S. advocates substantial reduction in domestic support and even elimination of export subsidies, yet on the premise that the developing WTO members must lower their market access thresholds, for example, to substantial reduce their tariffs and abolish quotas while opening their domestic market to developed WTO members, etc.

The WTO member countries (excluding the least developed countries) proposed to reduce tariffs by means of a tiered approach, that is, high tariffs are subject to more reductions. However, due to the different tariff structures between developed countries and developing countries, the choice of reduction means will pose a major challenge to negotiations. Meanwhile, as tariff is the only means for agricultural producers of developing countries to resist the subsidy protection of production and export by developed countries, the appropriate tariff reduction rates by developing countries also deserves attention. Besides, the treatment of non-tariff barriers (NTBs) was not clearly defined in the Doha Round Framework Agreement and is to be dealt within the negotiations. With the reduction of tariffs, NTBs have become increasingly apparent, and the WTO member countries are paying much more attention to NTBs, including the SPS, etc.

The abolishment of export subsidies and subsidies to products of special interests to developing countries is a major requirement for exporting countries of agricultural products and developing countries. Although developed countries promised to abolish all forms of export subsidies within a definite date, especially direct export subsidies, the specific date of abolishment and the detailed mode to ensure the abolishment remained to be negotiated. Meanwhile, in accordance with the Doha Round Framework Agreement, special and differentiated treatments would be granted to developing countries, least developed countries and net grain importing countries; developing members are allowed to abolish all forms of export subsidies in a longer period by phases; after the phased fulfillment of all disciplines and abolishment of all forms of export subsidies, developing countries will continue to enjoy the special and differentiated treatments stipulated in Article 9.4 of the Agreement on Agriculture. Whether this subsidy provision will be effective in the long term is likely to become a discussion topic in the next round of negotiations.

The Doha Round Framework Agreement decided to implement a tiered reduction of domestic support. That is to say, the higher trade-distorting degree domestic support has, the more reductions should be made. Nevertheless, the reduction basis required by the Framework Agreement is the obligatory level rather than the current actual support level, and the reduction from the obligatory level is usually higher than that from the current actual support level. Therefore, it should be further defined as to how to ensure the effective, substantial and gradual reduction of trade-distorting support provisions. In particular, whether the domestic support of developed countries regarding products of export interests for developing countries could be substantially reduced has become a major concern of developing countries. Besides, the Doha Round Framework Agreement retained the "Blue Box" measures (direct payments under production limiting programs), and allowed certain flexibility in the support ceiling of "Blue Box," that is, 5% of the total value of production of the agricultural product of a particular historical period. How this historical period should be determined remains to be discussed in negotiations. Moreover, in order to prevent the transfer of large scale of support measures into the new "Blue Box", member countries will seek for an agreement in the establishment of strict standards through negotiations.

3.3.3. *U.S.–China agricultural trade agreement*

Agricultural provisions in the U.S.–China Bilateral WTO Agreement (the Agreement) can be divided into three parts: the general rules of the Agricultural Agreement, specific agricultural product concession commitments and the Agreement on U.S.–China Agricultural Cooperation. These three parts proposed clear regulations on market access of agricultural products, domestic support, export subsidies, anti-dumping standards and the implementation standard of safeguard measures, etc.

✧ Market Access

Similar to other WTO members i.e. the U.S., the Chinese government also committed (1) to adopt a "tariff-only" regime on import of agricultural products and revoke all non-tariff measures which do not conform to the WTO rules; (2) to adopt a tariff rate quota (TRQ) system on sensitive commodities (i.e. bulk commodities); (3) to reduce import tariffs and increase TRQs; (4) to improve the transparency of quota management; (5) to realize the tariffication of some of the non-tariff measures on agricultural products; and finally, (6) to grant foreign enterprises with trade rights and distribution rights.

- Reducing tariffs, Increasing TRQs and Tariffication of Non-tariff Measures

China agreed to cut the import tariffs of agricultural products and gradually increase the TRQs of sensitive commodities. It was recommended that by 1 January 2004, the simple average tariff for agricultural products, which was at 22% in the preceding period, would be reduced to 17.5%. Tariffs on the U.S. priority products, which stood at 31% in the preceding period, would be reduced to 14%. The specific commitments are as follows: (1) to implement the TRQ system and the quasi-state monopoly system on cotton import; (2) to conduct tariff-only regime on import of dairy products; (3) to revoke all non-tariff measures inconsistent with the WTO rules; (4) to lower the import tariffs for fish from 25.3% during the preceding period, to 10.6% by 1 January 2005; (5) to implement the TRQ system and the quasistate monopoly system on import of cereals (excluding barley); (6) to carry out tariff-only regime on meat import and complete the tariff concession for meat import by 2004; (7) to implement tariff-only regime and cut the tariffs for import of oil seeds, soybean oil and specialized crops; (8) to reduce the tariff for wood and wood products from 10.6% during the preceeding period, to 3.8% by 1 January 2004.

- TRQ Management

China specially committed to manage TRQs with economic standards rather than political standards, to ensure the transparency and consistence of TRQ management, to formulate corresponding laws and regulations to ensure the smooth utilization of TRQs, and to adhere to the principle of full use of TRQs. If the original holders of TRQs fail to fully adhere to the TROs, they could transfer their TRQs to other possible trade entities of import. For example, TRQs held by state-owned trade enterprises (i.e. COFCO and Chinatex Corporation Limited) were not used in October that year and therefore were recommended to be transferred to nonstate-owned trade entities.

- Revocation of Non-tariff Measures Inconsistent with the WTO Rules

China promised to maintain the consistency of non-tariff measures (such as, SPS measures, domestic tax, etc.) with relevant WTO rules and to ensure the transparency and predictability of these measures so that they were built on sound scientific evidence rather than out of the need of politics or protectionism. An agreement was reached regarding the SPS standard of citrus, meat, wheat and other cereals in the Agreement on U.S.–China Agricultural Cooperation of the Agreement.

✧ Domestic Support and Export Subsidies

China has promised not to increase and to reduce trade-distorting domestic subsidies. The specific concession level is to be decided in accordance with the Agreement of the Multilateral Negotiation in Geneva and the Report of the WTO Working group. Besides, China will improve the transparency and predictability of domestic support measures. In terms of export subsidies, China promised to abolish all the export subsidies to agricultural products after entering the WTO.

✧ Trade Rights and Distribution Rights

Excluding the commodities in the specified product list (including wheat, corn, rice and cotton), China promised to gradually authorize all trade entities with trade rights within three years so that all the trade entities will have rights to import most products into any part of China. Commodities in the specified product list would rely on the import channel of state-owned trade entities. The Agreement also stipulated that China would gradually increase the number of commercial entities with import rights of these commodities so as to put an end to the state of import monopoly.

Within three years, China would gradually approve the participation of foreign enterprises in the distribution business of import commodities and allow them to provide a series of related distribution services, such as, maintenance, storage, stock, packaging, advertisement, transportation, shipping and express delivery, marketing, customer support, etc.

✧ Antidumping Standards and Product-specific Safeguard Provisions

In the Agreement, China has agreed that the current U.S. practice under its antidumping law with respect to non-market economy countries can apply to imports from China for 15 years after its accession; China also has agreed to a 12-year product-specific safeguard, which allows the U.S. to address rapidly increasing Chinese imports in a targeted fashion, if they are disrupting the U.S. market.

✧ Enhancement of Technical Cooperation and Science Exchange

China and the U.S. have agreed to enhance their cooperation and exchange in the high-tech field and to encourage the research and exploration by and the cooperation between the research institutes and agricultural enterprises in the high-tech field. The cooperation fields mainly include pasture and gardening products, biotechnology, meat, poultry, marine industry, natural resources and environment.

4. China's Food Trade and World's Food Trade

4.1. *The increasingly important position of China's food trade in world's food trade*

Since the 1960s, the increase of China's food output has gone hand-in-hand with the growth of world food output. Globally, food output grew by 2.7 times while China's food output went up by 3.4 times, obviously higher than the global average. The proportion of China's food output in the global food output also rose accordingly, increasing from 15.6% in the 1960s to over 20% with the highest proportion standing at one fourth. After deducting China's contribution in food output, the global food output increased by 2.4 times, far lower than the normal growth rate of China's food output. Besides, the ratio of China's food output to the total food output of other countries is even larger, rising from 18.4% to 32.6%. The proportion of China's food output in the global food output goes up continuously, and China is playing an increasingly important role.

In terms of grain varieties, China's rice output has been occupying a high proportion of the global output of rice, followed by China's corn output and wheat output. China's rice output accounted for around 30% of the global rice output, with the highest proportion standing at 37%; China's corn output made up 20% of global corn output, with the highest proportion sitting at 21.6%; China's wheat output took a stable proportion of 17% of the global wheat output, with the highest proportion standing at 20.1%.

In comparison with world trade, China's cereal trade did not occupy a high proportion of the global cereal trade. In terms of food export, China's food export volume went up by 16.8 times while the global food export volume increased by 3.4 times. It is clear that China's food output increased more rapidly. Correspondingly, the proportion of China's food export in the global food export also witnessed a fast growth, with the highest proportion standing at 8.4%. In terms of food import, China's food import volume rose by 4.0 times while the global food import volume went up by 3.4 times with a similar growth rate. The proportion of China's food import in the global food import stood between 6% and 7% with the highest proportion sitting at 8.5%. It can be seen that China exercised a sound regulation on food import. In terms of aggregate food trading volume, China's food trading volume increased by 6.5 times with the growth of export volume as the driving force, while the global food trading volume rose by 3.4 times with equal growth of import volume and export volume. The proportion of China's food trading volume in the global food trading volume basically stood at 5% and reached 8.5% in a few years. It is clear that China is occupying an increasingly important position in the global food trade in terms of import volume, export volume and aggregate trading volume. This result is closely related to the increase of China's food output.

According to statistics of 2007, China's food output stood at 0.5 billion tons, accounting for 23.6% of the global food output of 2.12 billion tons; the trading volume of staple agricultural products of China sat at 42.713 million tons, accounting for 13.4% of the global food trading volume of 0.32 billion tons; with a food import volume of 9.864 million tons and a food export volume of 1.554 million tons, China's net food import volume stood at

8.31 million tons; China's soybean output stood at 14 million tons while the global soybean output sat at 0.23 billion tons; China's soybean import stood at 30.82 million tons, accounting for 13.6% of the global soybean output, and its soybean export sat at 0.475 million tons. It is clear that the proportion of China's food output in the global food output is far higher than the proportion of China's food trading volume in the global food trading volume.

4.2. The correlation between China's food trade and the global food prices

The global food prices have experienced a rapid rise since 2006, which has been attributed by many people in the international community to "China factor". This saying, however, completely contradicts with China's food supply status. China is an important stabilizing factor rather than the driving force of the soaring food prices globally (Ding Shengjun, 2008). In sharp contrast to the decline of the global food output and reserve, China's food output went up continuously, leading to balanced food supply and demand, increased food reserve and ample market supply of food. China's food price merely experienced a mild and structural rise. From 2004 to 2007, China's aggregate food output went up in four consecutive years, increasing from 469.469 million tons in 2004 to 484.022 million tons in 2005, 497.499 million tons in 2006 and 501.50 million tons in 2007. China's food reserve has reached a rather high level, accounting for over 35% of China's annual food consumption and far higher than the minimum safety range of 17% to 18% proposed by FAO. Meanwhile, China has maintained a high food self-sufficiency rate which always stood above 95%. With a decreased import volume of cereals (excluding soybean) and an increased export volume of cereals, China has become a net exporter of cereals. China had a net export of cereals of 2.85 million tons in 2006, with the imported cereals being mostly used for the adjustment of grain varieties. On the other hand, the large quantities of soybean import by China greatly affected the global soybean prices. In the past decade, China's soybean consumption went up by 10 times with an annual soybean import of 30 million tons at present which equals to the soybean output of 20 million mu of cultivated land. China is in serious shortage of soybean supply and relies on export for 70% of its soybean supply. As the increase of China's soybean demand cannot be met by the increase of China's domestic soybean output, China will inevitably be associated with the global market. As long as China's soybean price goes hand-in-hand with the global food prices, other food prices in China will also go up along with the soybean price.

In terms of the correlation between international food prices and the net import volumes of different varieties of food from China, China's net import volume of food is directly proportional to international food prices. In terms of different varieties, the net import volume of soybean has the strongest correlation with international food prices with a correlation coefficient of 0.5 (soybean is included in the category of grain and oil rather than the category of grain), followed by rice with a correlation coefficient of 0.4, wheat with a correlation coefficient of 0.3 and corn with a correlation coefficient of 0.2. It can be seen that China's net import of grain exerts certain positive impacts on world food prices and that the increase of China's net import of food will promote the rise of world food prices yet with no large fluctuations, which can be proved by relevant figures. These facts indicate that the "large country effect" of China does exist. China should also undertake this responsibility.

Historical experience shows that food supply shortage will occur in China and the domestic food prices will have large fluctuations when China's food self-sufficiency rate stands below 95%. Nevertheless, China's food self-sufficiency rate remained above 95% in the past decade, realizing both the basic balance of domestic food supply and demand and fulfilling China's commitments in the World Food Summit in 1996.

4.3. *Elasticity of the trading volumes of different varieties of cereals in China*

Due to the "large country effect" of China's food trade, China's food trading volume exerts certain influence on the global food prices. This section studies the three major grain varieties, namely, rice, wheat and corn, as research objects to analyze the effects of their import volumes on their international prices. As the current international cereal price is calculated on the basis of the FOB price of a certain variety of this type of cereal, the regression analysis will have certain deficiencies. However, as different varieties of a certain type of cereal have certain complementarities, the current international price of this cereal can be used to indirectly reflect the prices of varieties which are not included in the calculation.

In terms of the trading volume, the international wheat price elasticity of China's wheat import volume is 2.0. That is to say, China's wheat import volume will rise by 2.0% whenever international wheat price goes up by 1%. From this we can see the feature of "purchase at high prices and sale at low prices" of China's wheat import and that China's food trade plays the role of repressing global food prices. The domestic wheat price elasticity of China's wheat importing is −0.2. This is to say, China's wheat import volume will decrease by 0.2% when domestic wheat price rises by 1%, indicating that the changes of domestic wheat prices do not exert significant influence on China's wheat import volume. China plays an important role in balancing world food production. China's wheat import price elasticity of the global wheat output stands at 7.2, while China's wheat import volume elasticity of China's wheat output is −6.2. It can be seen that China's wheat import volume is greatly affected by China's wheat output and that China's demand for wheat import will be substantially reduced once China's domestic wheat output satisfies China's domestic wheat demand. Meanwhile, due to the complementarity between China's rice import and wheat import, the global rice price elasticity of China's wheat import volume is 1.0. China's corn import does not have such features, indicating that corn is mostly used as feed grain and occasionally used for consumption. Corn import has certain uniqueness. First, China imports a small amount of corn. Second, China's corn import also plays the role of repressing global corn prices. With the decline in the global corn prices in the beginning of the 21st century, China's corn import volume also continued to fall, dropping to even less than 10,000 tons. The domestic price elasticity of China's corn import volume is 4.7, whereas the international price elasticity is 10.1.

The international rice price elasticity of China's rice import volume stands at merely 0.3. Though China's rice import volume is not large, the domestic rice price elasticity of China's rice import volume is 1.3. Therefore, China's rice import plays the role of repressing international food prices to a certain extent. Irrespective of fluctuations in domestic and international food prices, China will increase its rice import so as to fulfill its responsibilities as a large country. The international rice output elasticity of China's rice import is 4.0, implying that China increases its rice import when the international rice prices go up and decreases its rice import when the international rice prices go down. China's rice output elasticity of China's rice import is −3.1.

Therefore, the impact of China's food trade on the global cereal prices and China's "large country effect" or China's responsibilities as a large country are mainly manifested by the feature of China's import of cereals. Judging from the price elasticities of China's rice import volume, corn import volume and wheat volume, China's cereal import plays a certain role in repressing international food prices through a substantial reduction of its food import when the international food output falls. China exerts positive effects and soundly fulfills its responsibilities as a large country.

4.4. *Impacts of the changes of the dependence degree of China's food trade*

China' food trade and the global food trade influence each other. Globally, food prices will be affected if China does not export food. In the same way, international food prices will also be affected if China's food import volume exceeds the self-sufficiency rate of 95%.

In terms of rice, China's annual rice export stood at 1.959 million tons and its annual net export of rice sat at 1.584 million tons in the recent 10 years (1997–2006). If we assume that world cereal export volume is 30 million tons (the actual figure is smaller) and if China stops its rice export of 1.584 million tons which accounts for around 8% of the aggregate rice trading volume of the world, based on the international price elasticity of China's net import of rice of 0.3, it can be calculated that the international rice price will go up by 25%. In terms of wheat, as China's annual net export of wheat stood at 1.217 million tons which accounted for 1.1% of the global annual wheat trading volume, based on the above-mentioned elasticity, it can be calculated that wheat price will go up globally by 0.6% if China stops its export of wheat, which is not a big impact. In terms of corn, as China is a net importer of corn, no impact exists.

In the same way, if China's food import volume exceeds China's food self-sufficiency rate of 95%, then the international food prices will react. If we assume that China's aggregate cereal output is 0.4 billion tons, then 4 million tons of food will need to be imported when China increases its food import by 1%. The average proportions of rice, wheat and corn in China's aggregate food import in recent 10 years are 17%, 81% and 2% respectively. In this way, whenever China's food import volume surpasses the self-sufficiency rate of 95% by 1%, the international prices of rice and wheat will go up by 7.2% and 1.4% respectively, whereas the impact on international corn price will be negligible. If China continues to expand its food import volume until its food self-sufficiency rate falls to 93%, then the international prices of rice and wheat price will ascend by 14.5% and 2.88% respectively. If China's food self-sufficient rate drops to 90% and if the international food supply remains unchanged, then the prices of rice and wheat will rise internationally by 36% and 7% respectively. It is clear that changes in China's food self-sufficiency rates have the largest impact on world's rice market, the second largest impact on world's wheat market and negligible impact on world's corn market.

4.5. The correlation coefficient between the fluctuations of China's cereal output and the global cereal output

Though there is abundant food supply in the international food market in general, fluctuations do exist in the global food production between different years, thus affecting the availability of food supply in the international food market. That is why we continue to study the correlation coefficients between China's cereal output and world's cereal output fluctuations. Data shows that the real output of China and the world stood below the long-term trend in half of the investigated years, and above the long-term trend in the other half of the investigated years, with the same fluctuation direction in merely 10 years and a few two consecutive years and the opposite fluctuation directions in 60% of the investigated years.

In terms of different varieties of cereals, no significant linear correlation exists between the fluctuations in the production of China's major grain crops between different years and the fluctuations of international (excluding China) grain output between different years. The correlation coefficients of the fluctuations of rice and corn output are both negative, indicating that import and export of rice and corn could be economically profitable by making use of the reverse fluctuations of their output between different years and the consequent price differences. It should be noticed that the correlation coefficient of fluctuations of rice output is as high as 0.4. However, a further analysis shows that the available food supply of major grain producing countries constitutes the major part of the food supplies in world food market.

The correlation coefficients between the fluctuations of China's corn output between different years with the fluctuations of the corn outputs of five major grain exporting countries, namely, the U.S., Argentina, etc., between different years are all negative. Although the correlation did not reach the significant negative degree of "one decreases while the other increases", the non-correlation of the fluctuations in production implies that China will be able to obtain basically normal food supplies from grain exporting countries in the international market during years of poor harvests.

The correlation coefficients between the fluctuations of China's rice output with the fluctuations of the rice outputs of other major rice exporting countries in the world are all negative. The fluctuations of China's rice output between different years show certain complementarity with that of some of these countries, which is manifested by significant negative correlations. Therefore, there is no risk of supply availability in the international rice market.

The fluctuations in China's wheat output in comparison with the long-term trend have certain correlation with the fluctuations in the global wheat output although they did not obviously go hand-in-hand with the fluctuations in the global wheat output. The fluctuations in the wheat outputs of major wheat exporting countries and China indicate that obvious differences exist between the fluctuations in the wheat outputs of major wheat exporting countries and the fluctuations of China's wheat output and that their correlation coefficients stand below 0.3, belonging to the category of "non-correlation" in theory. The correlation coefficients among most of these wheat exporting countries are not high. The correlation coefficients between the fluctuations in the wheat outputs of major wheat exporting countries and the fluctuations in China's wheat output show that their correlation is small. Thus, it is highly unlikely for the major wheat importing countries to compete with China for food supplies in the international market in a particular year, and it's basically safe for China to ensure stable food supply in the domestic market by making use of the international wheat market.

Chapter 9

Measurement of Food Security — Food Gap

In this report, food security is defined as the ability to close the food gap. In the following two chapters, we will investigate the possibility of closing the food gap from both the domestic and international perspectives. This chapter focuses on the extent of China's food gap, namely, the difference between food production and food demand, and the relative number of food gap, namely, the ratio of food gap to food production and the ratio of food gap to food consumption because the amount of food gap alone can not reflect its ratio in food production and food consumption. In this chapter, we will first briefly analyze the current situation in China's food demand and supply, then analyze the absolute number and the relative number of domestic food gap over the years from statistical materials, and finally put forward the method used in this report to calculate food gap and the result of calculation.

1. The Current Situation of China's Food Demand and Supply

Since 2003, China's food output has gone up continuously to reach 501.5 billion kilograms in 2007, which is 70 billion excess kilograms higher than 430.7 billion kilograms, the ebb of food output in 2003. This was actually a correction of the relative surplus of food output and the decline of food stock in the previous seven to eight years. The outputs of wheat, paddy and corn have restored to 450 billion excess kilograms. Apart from the gap of soybean which needs to be made up by import, the consumption of wheat, paddy and corn stood between 425 billion to 435 billion kilograms, while domestic food consumption was around 510 kilograms. In 2007, China's domestic paddy output stood at around 185 billion kilograms, an increase of 26.5 billion excess kilograms over that of 2003, whereas domestic paddy consumption sat between 180 billion to 185 billion kilograms. As soybean and corn are major feed grain and industrial grain, this change led to the substantial increase in the consumption demand for soybean and corn. Data shows that China's corn consumption demand went up by 3.5% and its soybean consumption demand rose by 5.7% over the previous year in the year 2007–2008.

China's current food reserve is abundant. When inspecting agriculture and the spring ploughing in Hebei province in early April 2008, Chinese Premier Wen Jiabao said that China now had a food reserve of 0.15 to 0.2 billion tons, one time higher than the global average. Zhang Xiaoqiang, Deputy Director of National Development and Reform Commission (NDRC), revealed at the 6th Annual Conference of China Import & Export Enterprises on April 27, 2008 that the food reserve of Chinese government, enterprises and farmers totaled 500 billion Jin (about 0.25 billion tons), accounting about half of the total food consumption of the world in one year and far exceeding the minimum safe level of 18% of the ratio of food stock to food consumption regulated by the U.N.

China's food trade layout has changed slightly since 2005. In terms of import, China imported small amounts of wheat and paddy rice. China's wheat import volume decreased sharply from 3.54 million tons in 2005 to 0.61 billion tons in 2006 and 0.1 million tons in 2007. China's accumulated wheat import in the first five months of 2008 stood at merely 4,500 tons. Data showed that China's wheat import in May 2008 decreased by 76.9% over the same period of the previous year and 85.7% over the previous month. Nevertheless, China depended heavily on soybean import, with its accumulated soybean import in the first five months of 2008 standing at 13.65 million tons, 20.4% higher than that of the same period of the previous year. Table 9.1 showed that China imported 3.48 million tons of soybeans in May 2008, an increase of 17.6% over the same period of the previous year. In terms of export, China's accumulated rice import volume in the first five months of 2008 stood at 0.65 million tons; but China's monthly import of rice declined gradually since February 2008, sitting at merely 10,000 tons in May 2008, 80% below that of the same period of the previous year.

Table 9.1. China's Grain Import and Export in the First Half of 2008.

Unit: 10,000 tons

| Items | Year | | | | | | | | | May 2008 | |
	2005	2006	2007	2008	January	February	March	April	May	Year-on-year growth	Ring growth
Wheat import	354	61	10	0.45	0.15	0.07	0.06	0.07	0.01	−76.90%	−85.70%
Paddy rice import	52	73	49	22.2	10.2	3.6	4	3	1.4	−45.40%	−53.30%
Soybean import	2,659	2,827	3,082	1,365	344	202	232	239	348	17.60%	45.60%
Rice export	69	125	134	65	14	21	26	3	1	−80.00%	−66.70%
Corn export	864	310	492	11	3	3	2	3	0	−100.00%	−100.00%

Source: http://www.ndrc.gov.cn/zjgx/t20080618_218165.htm

2. China's Food Gap (The Gap between Food Production and Demand) over the Years in Accordance with Statistical Data

Generally speaking, the basic work of macro model research (such as, the equilibrium model of the agricultural sector or the general economic equilibrium model, etc.) is to create balance sheets of all the agricultural products involved in the models. In the so-called balance sheet, an equality relation of relevant historical data of investigated years in terms of the aggregate demand and the aggregate supply should be established. The balance sheet between the aggregate demand and the aggregate supply of a particular crop in a particular year is as follows:

Domestic Aggregate Output + Import = Demand + Export + (Year-End Stock of That Year – Start Stock of That Year)

As grain demand includes ration demand, feed demand, industrial grain demand, seed grain demand and the wastage, the above equation could be put as follows:

Domestic Aggregate Output + Import = (Consumption Demand of Residents + Feed Demand + Industrial Demand + Seed Demand + Wastage) + Export + (Year-End Stock of That Year – Start Stock of That Year)

Therefore, the gap between production and demand is:

Demand – Domestic Aggregate Output = (Import – Export) + (Year-End Stock of That Year – Start Stock of That Year)

Or:

(Consumption Demand of Residents + Feed Demand + Industrial Demand + Seed Demand + Wastage) – Domestic Aggregate Output = (Import–Export) + (Year-End Stock of That Year – Start Stock of That Year)

Therefore, food gap could be obtained from two methods, namely, the difference between food demand and food output on the left of the equation, or the sum of net import and the decrease of food stock of that year on the right of the equation. The feasibility of the two methods will be discussed in the following paragraphs. The data presented here has been derived from domestic statistical materials or articles of scholars as well as the Department of Agriculture of the U.S.

2.1. *Method I: Difference between food demand and food output*

In this method, we need to obtain data of food output and food demand to calculate the amount of food gap. The data of food output could be easily ascertained. Similarly, food consumption could be deduced in accordance with per capita food consumption and the total population. Estimation or trend deduction and estimation could be made on feed demand, industrial demand, seed demand and the wastage on the basis of certain sporadic materials and data.

The following figure shows that China's annual food output presented a rising trend in the past 30 years of reform and opening-up, increasing from 0.3 billion tons in the last century to 0.5 billion tons in the late 1990s with several large fluctuations and basically complying with the rule of "two years of output increase and one year of output decrease". China's annual food output reached the peak at the end of last century, basically standing at 0.5 billion tons from 1996 to 1999, and then decreased continuously in the beginning of the 21st century to the bottom of 0.43 billion tons in 2003. Nevertheless, China's annual food output started to pick up in recent years and restored to 0.5 billion tons.

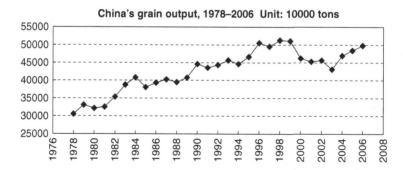

Figure 9.1. China's grain output, 1978–2006.
Source: China Grain Development Report 2007, Nie Zhenbang (ed.). Please see attached Table 9.2 for detailed data (method I).

Table 9.2. China's Food Gap Data I

Year	Food output (10,000 tons)	Food demand (10,000 tons)	Food gap (10,000 tons)	Ratio of food gap to food output	Ratio of food gap to food demand
1978	30,477	30,462	15	0.05%	0.05%
1979	33,212	31,946	1,266	3.81%	3.96%
1980	32,056	33,023	−967	−3.02%	−2.93%
1981	32,502	33,665	−1,163	−3.58%	−3.45%
1982	35,450	34,426	1,024	2.89%	2.97%
1983	38,728	34,921	3,807	9.83%	10.90%
1984	40,731	35,841	4,890	12.01%	13.64%
1985	37,911	37,186	725	1.91%	1.95%
1986	39,151	38,018	1,133	2.89%	2.98%
1987	40,298	38,439	1,859	4.61%	4.84%
1988	39,408	39,141	267	0.68%	0.68%
1989	40,755	39,800	955	2.34%	2.40%
1990	44,624	40,872	3,752	8.41%	9.18%
1991	43,529	42,010	1,519	3.49%	3.62%
1992	44,265.8	42,891	1,374.8	3.11%	3.21%
1993	45,648.8	43,842	1,806.8	3.96%	4.12%
1994	44,510	44,859	−349	−0.78%	−0.78%
1995	46,662	45,302	1,360	2.91%	3.00%
1996	50,450	45,927	4,523	8.97%	9.85%
1997	49,417	46,087	3,330	6.74%	7.23%
1998	51,229.5	46,557	4,672.5	9.12%	10.04%
1999	50,839	47,013	3,826	7.53%	8.14%
2000	46,218	47,926	−1,708	−3.70%	−3.56%
2001	45,264	48,093	−2,829	−6.25%	−5.88%
2002	45,706	8,453	−2,747	−6.01%	−5.67%
2003	43,070	48,625	−5,555	−12.90%	−11.42%
2004	46,947	49,090	−2,143	−4.56%	−4.37%
2005	48,402.19049	49,775	−1,372.81	−2.84%	−2.76%
2006	49,747.9	50,800	−1,052.1	−2.11%	−2.07%

Note: Calculated in accordance with the formula: Food Gap = Food Output − Food Demand.

Source: Data of food demand from 2003–2006 is from China Grain Development Report; data of food demand from 1996–2002 is from China Grain Security Report released by the Ministry of Agriculture of China in May 2004; data of 1995 was calculated in accordance with page 85 of the Research on China's Food Security (Xiao Guoan, 2005); data of food demand from 1978 to 1994 is form Ye Xingqing (1999).

Food demand includes ration demand, feed demand, industrial demand, seed demand and the wastage. The different types of demand are not taken into account in this report. The statistics of domestic food demand was found in statistical reports until in recent years, such as, the data of 2003 and 2006 come from China Grain Development Report over the years respectively. Older data is sought from two channels: data from 1978 to 1994 is taken from from Ye Xingqing (1999),[1] and data from 1996 to 2002 is taken from China Grain Security Report released by the Ministry of Agriculture of China in May 2004. From the changing trends in China's food demand in Fig. 9.2, it can be seen that the annual food demand showed a increasing trend year-by-year since the implementation of reform and opening-up in 1978, rising slowly from 0.3 billion tons to 0.5 billion tons.

The absolute number and the relative number (the ratio of food gap to food output and the ratio of food gap to food consumption) of food gap could be calculated on the basis of the data of food demand in the below figure (Figs. 9.3 and 9.4). It can be seen that China's food output outstripped

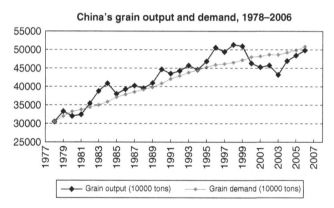

Figure 9.2. China's Grain Output and Demand, 1978–2006. Please See Attached Table 9.2 for Detailed Data (Method I).

---

[1]The summarized data of this article came from the Research on the Balance of China Grain Demand and Supply written by the Research Team of the Agricultural Investigation Group of the Agricultural Economy Department of the State Development and Planning Commission and National Bureau of Statistics in 1995, and the Significant Changes of China's Grain Demand and Supply Relationship since the Reform and Opening-up and More Efforts Need to be Made to Ensure the Basic Balance of Grain Supply and Demand written by the Circulation Office of the Department of Trade & External Economic Relations Statistics of National Bureau of Statistics in 1998.

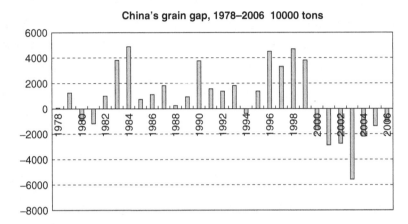

Figure 9.3. China's Grain Gap, 1978–2006. Please See Attached Table 9.2 for Detailed Data (Method I).

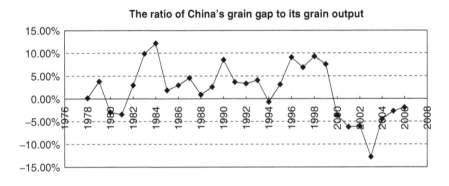

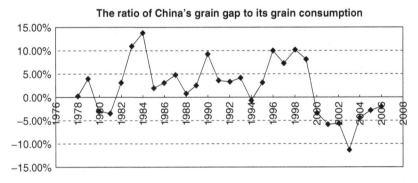

Figure 9.4. The Ratios of China's Grain Gap to its Grain Output and Grain Consumption, 1978–2006. Please See Attached Table 9.2 for Detailed Data (Method I).

its food demand in most of the past 30 years and that China's food surplus reached the peak at 48.9 million tons in 1984. Since 2000, China's food output has dropped continuously while its food demand had maintained a steady growth. Therefore, food gap occurred in several consecutive years, among which the food gap of 2003 reached the peak at 55.55 million tons. The changing trend in the ratio of food gap to food output was basically consistent with that of food gap. The ratio of China's food gap to its food output stood at 12.9% and the ratio of China's food gap to its food consumption sat at 11.4% in 2003. It should be noted that the major reason for China's food gap reaching a historical record of 55 million excess tons in 2003 was that China's food output increased continuously in the 1990s with regional and structural food surplus and then started a decline for several consecutive years since 1999 to reach 430.7 million tons in 2003, the lowest in the past decade. The state adopted many preferential agricultural policies after 2004. As a result, China's food output started to increase which further narrowed the food gap.

2.2. *Method II*: *The sum of net import and the changes in food stock*

According to the formula:

$$\text{Food Gap} = \text{Net Food Import} + \text{Decrease of Food Stock}$$
$$\text{The Ratio of Food Gap} = \text{Food Gap/Food Output}$$
$$= (\text{Net Food Import} + \text{Decrease of Food Stock})/\text{Food Output}$$

Data of net food import was obtained from China's Foreign Economic Relations and Trade. Please see attached Tables 9.3 and 9.4 for detailed data. Nevertheless, data of food stock is hard to be obtained. Food stock at a specific time point includes two aspects: food stock of rural households and food stock of enterprises. The central food reserve, local food reserve and food reserve of state-owned enterprises are all included in food stock of enterprises. Food stock of rural households is mostly fine grain and new grain of the present year, which could basically meet farmers' demand for ration in half-a-year at present. Nevertheless, food stock of rural households

Table 9.3.   China's Food Import and Export Over the Years.

Year	Aggregate food export (10,000 tons)	Aggregate food import (10,000 tons)	Net food import (10,000 tons)
1950	122.58	6.69	−115.89
1951	197.11	—	−197.11
1952	152.88	0.01	−152.87
1953	182.62	1.46	−181.16
1954	171.1	3	−168.1
1955	223.34	18.22	−205.12
1956	265.12	14.92	−250.2
1957	209.26	16.68	−192.58
1958	288.34	22.35	−265.99
1959	415.75	0.2	−415.55
1960	272.04	6.63	−265.41
1961	135.5	580.97	445.47
1962	103.09	492.3	389.21
1963	149.01	595.2	446.19
1964	182.08	657.01	474.93
1965	241.65	640.52	398.87
1966	288.5	643.78	355.28
1967	299.44	470.19	170.75
1968	260.13	459.64	199.51
1969	223.75	378.63	154.88
1970	211.91	535.96	324.05
1971	261.75	317.32	55.57
1972	292.56	475.62	183.06
1973	389.31	812.79	423.48
1974	364.39	812.73	448.34
1975	280.61	373.5	92.89
1976	176.47	236.65	60.18
1977	165.7	734.48	568.78
1978	187.72	883.25	695.53
1979	165.08	1,235.53	1,070.45
1980	161.83	1,342.93	1,181.1
1981	126.08	1,481.22	1,355.14
1982	125.12	1,611.69	1,486.57
1983	196.13	1,343.51	1,147.38
1984	319	1,041	722
1985	932	600	−332
1986	942	773	−169
1987	737	1,628	891
1988	717	1,533	816
1989	656	1,658	1,002

(*Continued*)

Table 9.3. (*Continued*)

Year	Aggregate food export (10,000 tons)	Aggregate food import (10,000 tons)	Net food import (10,000 tons)
1990	583	1,372	789
1991	1,086	1,345	259
1992	1,364	1,175	−189
1993	1,365.1	742.5	−622.6
1994	1,187.5	924.9	−262.6
1995	102.5	2,070.1	1,967.6
1996	143.6	1,195.5	1,051.9
1997	853.6	705.5	−148.1
1998	906.5	708.6	−197.9
1999	759	772.1	13.1
2000	1,401.3	1,356.8	−44.5
2001	903.1	1,738.4	835.3
2002	1,514.3	1,416.7	−97.6
2003	2,279.3	2,288.1	8.8
2004	434.4	2,993.1	2,558.7
2005	998.5	3,280	2,281.5
2006	584.7	3,178.9	2,594.2

Source: *China's Foreign Economic Relations and Trade over the years.*

is dispersed. In addition, the scale of rural household is small and the number of rural households is large. Therefore, it is difficult to calculate the exact amount of food stock of rural households.

The present available time series data includes the amount of grain procurement by state-owned enterprises released by state grain departments and data of different grain varieties of China from the Department of Agriculture of the U.S. Lu Feng and Xie Ya (2007) believed that the procurement amount of grains by state-owned enterprises did not equal to but was associated with the change in grain stock. Based on the net increase of grain stock calculated by deducting the amount of grain sales from the amount of grain procurement in combination with the data of grain stock from the Department of Agriculture of the U.S., we can obtain the ceiling of China's present grain stock.

Therefore, we adopt a middle course when investigating grain stock, that is, to consider the difference between the amount of grain sales and the amount of grain procurement as a rough net decrease of grain stock.

Table 9.4.   Import and Export of Different Grain Varieties of China Over the Years.

Year	Wheat export (10,000 tons)	Wheat import (10,000 tons)	Rice export (10,000 tons)	Rice import (10,000 tons)	Corn export (10,000 tons)	Corn import (10,000 tons)	Soybean export (10,000 tons)	Soybean import (10000 tons)
1950			4.93	5.71			91.35	
1951			12.66				100.89	
1952			33.48				86.47	0.01
1953		1.36	56.12				92	0.01
1954		2.69	54.02				90.69	0.01
1955		2.16	70.03	15.72			105.81	
1956		2.26	167.7	11.47		1.2	112.43	
1957		4.99	52.94	10.6		1.09	114.11	
1958		14.83	139.71	3.16		4.19	122.44	
1959			177.35			0.2	172.67	
1960		3.87	107.15	2.76			111.07	
1961	358.17		42.83	36.3		8.62	40.88	
1962	353.56		45.79	17.41		49.63	25.92	
1963	558.77		68.45	10.02		21.76	40.92	
1964	536.87		76.16	15.91		28.15	58.99	
1965	607.27		98.49	18.08		9.85	65.32	
1966	621.38		148.74	19.86		2.53	65.09	
1967	439.46		157.66	9.47		21.05	67.03	
1968	445.14		129.92	3.92		10.24	68.8	
1969	374.02		117.85	1.71		2.9	59.45	
1970	530.21		127.95	4.2		1.3	47.01	
1971	302.2		129.2	13.1		2.02	58.79	
1972	433.36		142.57	19.7		22.33	41.21	
1973	629.85		263.08	6.67		160.56	39.95	12.79
1974	538.34		206.05	12.07		190.53	47.14	71.19
1975	249.12		162.96	7.02		13.68	40.47	3.68
1976	202.19		87.61	29.02		2.39	19.97	2.95
1977	687.58		103.29	12.53		0.77	12.95	33.07
1978	766.73		143.52	17.07		79.4	11.29	19.03
1979	870.98		105.31	12.39		279.16	30.59	57.95
1980	1,097.17		111.64	14.76		169.63	11.35	53.39
1981	1,307.01		58.33	19.64		74.79	13.6	56.47
1982	1,353.43		45.71	39.59		161.15	12.69	33
1983	1,101.91		56.59	16.08	6	198.63	33.39	33.4
1984	1,000		118.9	24.6	95	6	83.4	0
1985	541		101.9	31.3	633.7	9.1	115.1	0.1
1986	611		95.7	31.9	564	58.8	130.1	32.2
1987	1,320		98.9	48.6	392	154.2	171.4	42.8

(*Continued*)

Table 9.4. (*Continued*)

Year	Wheat export (10,000 tons)	Wheat import (10,000 tons)	Rice export (10,000 tons)	Rice import (10,000 tons)	Corn export (10,000 tons)	Corn import (10,000 tons)	Soybean export (10,000 tons)	Soybean import (10000 tons)
1988	0.7	1,454.7	70.5	33.2	391.2	10.9	145.9	3.3
1989	0.1	1,488	33.9	101.2	350.1	6.8	117.1	0.1
1990	0.3	1,252.7	30.3	4.9	340.4	36.9	91	0.1
1991	0.2	1,236.7	69.2	4.6	778.2	0.1	106.5	0
1992	0.3	1,058.1	120.6	10.4	1034	0	84.5	12.1
1993	8.7	642.3	170.9	9.6	1109.7	0	34.5	9.9
1994		730	177.3	51.3	874	0.1	92.7	5.2
1995	1.6	1,158.6	6.7	162.9	11.3	518.1	37.5	29.4
1996		824.6	26.4	76.1	15.9	44.1	19.2	110.8
1997	0.1	186.1	93.7	32.6	661.7	0	19	279
1998	0.6	148.9	374.5	24.4	468.6	25.1	17	320
1999	0.1	44.8	270.3	16.8	430.5	7	20.4	431.9
2000	0.3	87.6	294.8	23.9	1,046.6	0	21.1	1041.9
2001	45.5	69	184.7	26.9	599.8	3.6	24.8	1394
2002	68.7	60.4	196.5	23.6	1,167.3	0.6	27.6	1131.5
2003	223.7	42.4	260.1	25.7	1,639.9	0	26.7	2074.1
2004	78.4	723.3	91	76.1	232		33	2023
2005	26	351	68.5	52	864		40	2659
2006	111.4	58.4	125	73	310	6.5	38.3	2827

Source: China's Foreign Economic Relations and Trade over the years.

We also assume that the annual grain stock of rural households witnessed steady changes. Although grain stock of rural households might show a declining trend with the continuous improvement of marketization of grain market and the income increase of farmers, its changes should generally be rather gentle. Therefore, grain stock of rural households is omitted when we investigate the changes of grain stock.

Besides, this report also used data from the Department of Agriculture of the U.S. to calculate food gap so as to compare with the above-mentioned data.

It should be noticed that the data of net grain import and the decrease of annual grain stock was calculated in terms of trade grain while the data of grain output was calculated in terms of raw grain. Although the amount of raw grain and the amount of trade grain could be conversed, we did not

Food Security and Farm Land Protection in China

Table 9.5. Grain Procurement of State-owned Enterprises, 1978–2006.

(Trade Grain) unit: 10,000 tons

Year	Aggregate grain procurement	Procurement of different grain varieties			
		Wheat	Rice	Corn	Soybean
1978	5,110.2	1,176.8	1,955.9	1,046.7	216.0
1979	5,925.0	1,562.5	2,201.0	1,281.0	205.0
1980	5,882.1	1,396.1	2,214.5	1,357.8	296.5
1981	6,250.5	1,418.3	2,421.1	1,408.0	412.6
1982	7,367.5	1,933.6	2,900.3	1,427.4	401.7
1983	9,879.6	2,763.3	3,312.4	2,337.8	409.8
1984	11,165.9	3,427.0	3,858.1	2,588.1	382.4
1985	7,925.0	2,666.1	3,012.9	1,365.8	503.3
1986	9,453.3	2,842.2	3,258.7	1,871.1	653.7
1987	9,920.1	2,816.2	3,143.7	2,848.6	609.7
1988	9340.4	2,673.9	3,185.9	2,414.7	693.5
1989	10,040.2	2,855.5	3,622.9	2,587.7	620.0
1990	12,364.5	3,646.6	4,316.0	3,072.8	661.2
1991	11,423.0	3,392.5	3,810.0	3,338.4	582.2
1992	10,414.4	3,841.4	3,272.6	2,621.7	406.1
1993	9,234.0	3,373.1	2,505.0	2,470.0	606.2
1994	9,226.4	3,230.4	2,697.6	2,185.0	732.2
1995	9,443.8	3,125.0	3,061.4	2,435.6	522.5
1996	11,918.3	3,614.8	3,382.2	4,224.7	437.8
1997	11,535.5	4,600.3	3,510.6	2,692.2	515.2
1998	9,654.5	2,795.6	2,562.0	3,867.4	351.0
1999	12,807.7	3,863.3	3,186.1	5,425.0	246.6
2000	11,695.1	4,018.2	3,327.3	4,019.2	237.9
2001	11,783.9	4,440.8	2,796.6	4,128.1	326.4
2002	10,826.3	4,201.3	2,189.6	4,182.0	140.4
2003	9,718.9	3,677.0	2,110.1	3,704.9	124.4
2004	8,920.0	3,449.0	2,138.0	3,159.0	41.0
2005	11,493.9	3,745.3	2,572.3	4,529.8	506.0
2006	12,257	6,040	2,153	3,425	492

Source: China Grain Development Report 2006. Data of 2006 comes from the website of State Administration of Grain.

conduct conversion between the two because the focus of our investigation was the amount of food gap and the ratio of food gap to food output and the changes in the ratio could also illustrate this problem. Figures 9.5–9.7 show the data of net grain import and the difference between grain procurement

Table 9.6. Grain Sales of State-owned Enterprises, 1978–2006.

(Trade Grain) unit: 10,000 tons

Year	Aggregate grain sales	Sales of different grain varieties			
		Wheat	Rice	Corn	Soybean
1978	5,343.5	1,869.5	1,774.0	876.0	162.5
1979	5,679.1	1,940.5	1,826.0	1,068.0	180.0
1980	6,416.8	2,257.0	2,014.5	1,301.5	204.5
1981	7,223.3	2,563.5	2,123.0	1,622.5	239.0
1982	7,710.4	2,858.0	2,289.5	1,596.5	272.0
1983	8,003.2	3,006.0	2,497.5	1,458.5	289.0
1984	10,417.9	3,699.5	3,438.5	1,932.0	355.5
1985	8,564.9	3,078.5	3,006.5	1,309.5	323.0
1986	9,347.7	3,618.0	3,244.0	1,357.0	321.5
1987	9,190.8	3,643.5	3,080.0	1,424.0	355.5
1988	10,091.0	3,885.0	121.4	1,898.5	406.5
1989	8,931.1	3,522.0	2,566.0	1,846.0	346.5
1990	9,033.3	3,575.0	2,770.5	1,723.0	341.5
1991	10,433.0	4,085.0	3,267.5	2,065.0	383.5
1992	9,000.0	3,481.5	3,046.0	1,637.5	257.5
1993	6,700.3	2,848.5	2,128.5	1,088.0	230.0
1994	7,648.4	3,328.0	2,609.5	1,121.5	234.0
1995	9,264.2	3,707.5	2,897.0	1,570.0	620.5
1996	7,340.9	3,090.5	2,259.5	1,346.5	357.0
1997	6,830.7	2,439.5	2,043.0	1,632.5	429.5
1998	6,116.0	2,137.0	1,795.5	1,648.5	348.5
1999	9,353.3	3,137.0	2,421.0	3,197.5	439.5
2000	12,556.9	3,962.0	3,030.0	4,718.5	645.5
2001	8,528.6	3,225.5	2,154.0	2,577.0	439.0
2002	12,070.1	4,733.0	3,155.5	3,551.5	510.5
2003	13,327.6	5,458.5	3,520.5	3,817.0	361.0
2004	11,944.0	4,641.0	3,247.0	3,575.0	310.0
2005	12,138.4	4,277.1	2,556.7	4,348.7	841.7
2006	12,034	4,246	2,671	4,133	848

Source: *China Grain Development Report 2006. Data of 2006 comes from the website of State Administration of Grain.*

and sales by state-owned enterprises as well as the changing trends in food gap.

In accordance with Method II, namely, to calculate the amount of food gap by adding net food import and the decrease of food stock, we can derive that the largest food gap is 48.520 million tons in 1997 and the largest food

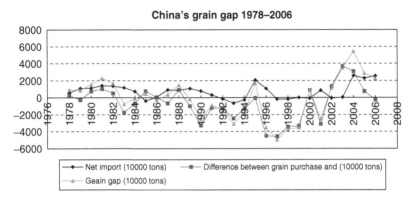

Figure 9.5. Amount of China's Grain Gap, 1978–2006. Please See Attached Table 9.7 for Detailed Data (Method II).

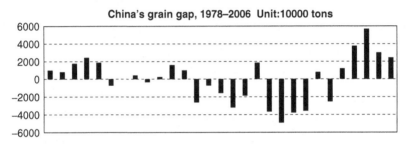

Figure 9.6. China's Grain Gap, 1978–2006. Please See Attached Table 9.7 for Detailed Data (Method II).

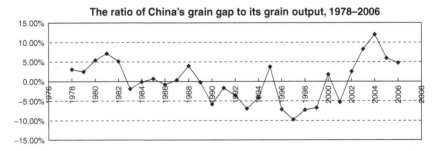

Figure 9.7. The Ratio of China's Grain Gap to its Grain Output, 1978–2006. Please See Attached Table 9.7 for Detailed Data (Method II).

Table 9.7. China's Food Gap Data II.

Year	Net food import (10,000 tons)	Difference between food procurement and sales	Food gap (10,000 tons)	Food output (10,000 tons)	Ratio of food gap to food output (%)
1978	695.53	233.3	928.83	30,477	3.05
1979	1,070.45	−245.9	824.55	33,212	2.48
1980	1,181.1	534.7	1,715.8	32,056	5.35
1981	1,355.14	972.8	2,327.94	32,502	7.16
1982	1,486.57	342.9	1,829.47	35,450	5.16
1983	1,147.38	−1,876.4	−729.02	38,728	−1.88
1984	722	−748	−26	40,731	−0.06
1985	−332	639.9	307.9	37,911	0.81
1986	−169	−105.6	−274.6	39,151	−0.70
1987	891	−729.3	161.7	40,298	0.40
1988	816	750.6	1,566.6	39,408	3.98
1989	1,002	−1,109.1	−107.1	40,755	−0.26
1990	789	−3,331.2	−2,542.2	44,624	−5.70
1991	259	−990	−731	43,529	−1.68
1992	−189	−1,414.4	−1,603.4	44,265.8	−3.62
1993	−622.6	−2,533.7	−3,156.3	45,648.8	−6.91
1994	−262.6	−1,578	−1,840.6	44,510	−4.14
1995	1,967.6	−179.6	1,788	46,662	3.83
1996	1,051.9	−4,577.4	−3,525.5	50,450	−6.99
1997	−148.1	−4,704.8	−4,852.9	49,417	−9.82
1998	−197.9	−3,538.5	−3,736.4	51,229.5	−7.29
1999	13.1	−3,454.4	−3,441.3	50,839	−6.77
2000	−44.5	861.8	817.3	46,218	1.77
2001	835.3	−3,255.3	−2,420	45,264	−5.35
2002	−97.6	1,243.8	1,146.2	45,706	2.51
2003	8.8	3,608.7	3,617.5	43,070	8.40
2004	2,558.7	3,024	5,582.7	46,947	11.89
2005	2,281.5	644.5	2,926	48,402	6.05
2006	2,594.2	−223	2,371.2	49,747.9	4.77

Note: Calculated in accordance with the formula: Food Gap = Food Output − Food Demand.

surplus is 55.827 million tons in 2004. China's food supply and demand tended to equalize in the long-term. However, the changing trend in the food gap obtained through this method is very different from that of Method II. Therefore, obvious deficiencies exist in this data.

2.3. *The gap in food demand and supply calculated by using data of the department of agriculture of the U.S.*

The Production, Supply and Distribution Online Database[2] of the Department of Agriculture of the U.S. includes rather comprehensive data about the output, consumption, opening and closing stock, and import of different varieties of agricultural products. However, as this database does not include the data of food stock of rural households, the difference between domestic food production and domestic food consumption is used as food gap in this report so that the obtained data is closer to the actual data.

Besides, a difference also exists in the caliber of grain between China and other countries. That is, China's statistics of grain includes soybean while that of other countries does not. According to the grain-caliber of other countries, grains include potatoes and cereals, and cereals include merely paddy, wheat, corn and other roughage. In this report, we use the sum of the domestic output of rice, corn, wheat and soybean and the sum of the domestic consumption of the four varieties as the substitutes of the aggregate food output and the aggregate food consumption, respectively.

From the above figure, it can be seen that the trends in China's food output and its food consumption reflected by data from the Department

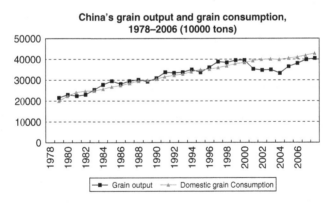

Figure 9.8. China's Grain Output and Consumption, 1978–2006.
Source: *USDA* http://www.fas.usda.gov/psdonline/psdQuery.aspx. Please see attached Table 9.8 for detailed data (method III).

---

[2]http://www.fas.usda.gov/psdonline/psdQuery.aspx

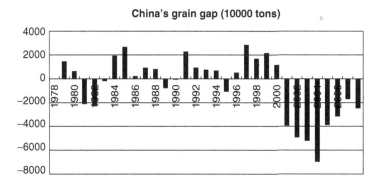

Figure 9.9.   China's Grain Gap, 1978–2006.
Source: *USDA* http://www.fas.usda.gov/psdonline/psdQuery.aspx. Please see attached
Table 9.8 for detailed data (method III).

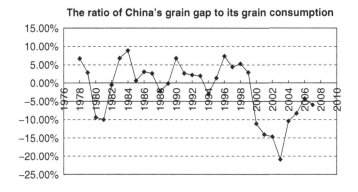

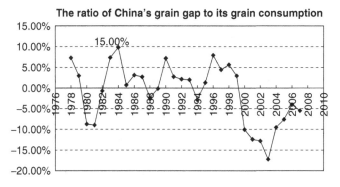

Figure 9.10.   The Ratios of China's Grain Gap to its Grain Output and Grain Consumption,
1978–2006.
Source: *USDA* http://www.fas.usda.gov/psdonline/psdQuery.aspx. Please see attached
Table 9.8 for detailed data (method III).

Table 9.8. China's Food Gap Data III.

Year	Corn Production (1000 MT)	Corn Domestic Consumption (1000 MT)	Oilseed, Soybean Production (1,000 MT)	Oilseed, Soybean Domestic Consumption (1,000 MT)	Rice, Milled Production (1,000 MT)	Rice, Milled Domestic Consumption (1,000 MT)	Wheat Production (1,000 MT)	Wheat Domestic Consumption (1,000 MT)	Food output (1000 MT)	Domestic food consumption (1000 MT)	Food gap (1000 MT)	Food Gap/Food Production
1978	55,945	48,600	7,565	7,552	95,850	89,868	53,840	52,887	213,200	198,907	14,293	6.70%
1979	60,035	53,300	7,460	8,063	100,625	96,527	62,730	66,595	230,850	224,485	6,365	2.76%
1980	62,600	61,800	7,940	8,337	97,934	98,587	55,210	75,999	223,684	244,723	−21,039	−9.41%
1981	59,205	62,000	9,325	9,745	100,768	101,085	59,640	78,840	228,938	251,670	−22,732	−9.93%
1982	60,560	61,400	9,030	8,740	113,117	103,350	68,470	79,470	251,177	252,960	−1,783	−0.71%
1983	68,205	61,800	9,760	8,960	118,206	105,212	81,390	82,990	277,561	258,962	18,599	6.70%
1984	73,410	61,200	9,695	8,615	124,779	110,461	87,815	89,105	295,699	269,381	26,318	8.90%
1985	63,826	59,700	10,509	9,529	117,999	111,894	85,810	95,155	278,144	276,278	1,866	0.67%
1986	70,856	64,100	11,614	10,054	120,557	112,685	90,040	97,265	293,067	284,104	8,963	3.06%
1987	79,240	67,300	12,184	10,910	121,716	115,939	87,764	99,040	300,904	293,189	7,715	2.56%
1988	77,351	69,000	11,645	10,469	118,377	118,605	85,432	101,826	292,805	299,900	−7,095	−2.42%
1989	78,928	74,200	10,227	9,121	126,091	120,822	90,807	102,367	306,053	306,510	−457	−0.15%
1990	96,820	79,850	11,000	9,713	132,532	123,911	98,229	102,598	338,581	316,072	22,509	6.65%
1991	98,770	83,200	9,710	8,756	128,667	126,827	96,000	105,429	333,147	324,212	8,935	2.68%
1992	95,380	87,800	10,300	10,150	130,354	128,135	101,590	104,281	337,624	330,366	7,258	2.15%
1993	102,700	92,900	15,310	14,335	124,390	129,340	106,390	105,343	348,790	341,918	6,872	1.97%

(*Continued*)

Table 9.8. (*Continued*)

Year	Corn		Oilseed, Soybean		Rice, Milled		Wheat		Food output (1,000 MT)	Domestic food consumption (1,000 MT)	Food gap (1,000 MT)	Food Gap/Food Production
	Production (1,000 MT)	Domestic Consumption (1,000 MT)	Production (1,000 MT)	Domestic Consumption (1,000 MT)	Production (1,000 MT)	Domestic Consumption (1,000 MT)	Production (1,000 MT)	Domestic Consumption (1,000 MT)				
1994	99,280	97,000	16,000	15,761	123,151	130,117	99,300	105,355	337,731	348,233	−10,502	−3.11%
1995	112,000	101,200	13,500	14,073	129,650	131,237	102,215	106,499	357,365	353,009	4,356	1.22%
1996	127,470	105,750	13,220	14,309	136,570	131,954	110,570	107,615	387,830	359,628	28,202	7.27%
1997	104,309	109,500	14,728	15,472	140,490	132,700	123,289	109,056	382,816	366,728	16,088	4.20%
1998	132,954	113,920	15,152	19,929	139,100	134,100	109,726	108,250	396,932	376,199	20,733	5.22%
1999	128,086	117,300	14,290	22,894	138,936	134,200	113,880	109,340	395,192	383,734	11,458	2.90%
2000	106,000	120,240	15,400	26,697	131,536	134,300	99,640	110,278	352,576	391,515	−38,939	−11.04%
2001	114,088	123,100	15,410	28,310	124,306	136,500	93,873	108,742	347,677	396,652	−48,975	−14.09%
2002	121,300	125,900	16,510	35,290	122,180	135,700	90,290	105,200	350,280	402,090	−51,810	−14.79%
2003	115,830	128,400	15,394	34,375	112,462	132,100	86,490	104,500	330,176	399,375	−69,199	−20.96%
2004	130,290	131,000	17,400	40,212	125,363	130,300	91,950	102,000	365,003	403,512	−38,509	−10.55%
2005	139,365	137,000	16,350	44,440	126,414	128,000	97,450	101,500	379,579	410,940	−31,361	−8.26%
2006	151,600	145,000	15,200	45,397	127,200	127,200	108,470	102,000	402,470	419,597	−17,127	−4.26%
2007	151,830	149,000	13,500	48,850	129,840	127,340	109,860	104,000	405,030	429,190	−24,160	−5.96%

Note: MT represents metric ton. 1 MT is considered as roughly equal to 1 ton.

Source: USDA of the Department of Agriculture of the U.S. http://www.fas.usda.gov/psdonline/psdQuery.aspx

of Agriculture of the U.S. complies with the trend calculated by using Method I. The only difference is that the data of the aggregate food output and consumption are lower than the domestic statistics because we use the sum of rice, corn, wheat and soybean as a substitute of the total amount of food. In accordance with this method, China's food gap also peaked in 2003, whereas the ratio of China's food gap to its food consumption stood as high as 17.3%, the highest among the calculations by using the three methods.

In conclusion, due to the availability of data, the difference between China's food output and demand could reflect China's food gap rather faithfully. In addition, as the trend of food gap reflected by domestic statistics complies with that reflected by data of the Department of Agriculture of the U.S., we consider the first set of data as of high accuracy and the main basis for our calculation in the following passage also comes from the first set of data.

3. The Method Used in this Report

The above passage shows the absolute number and the relative number of food gap calculated in accordance with both domestic and foreign statistical data. In this report, we propose a more accurate method. That is, to quantize the cases of "once in 100 years", "once in 50 years" and "once in 10 years" as the probabilities of 1%, 2% and 10% of the occurrence of food shortage, and then to calculate the ratios of food gap to food output or consumption when food shortage happens at the probabilities of 1%, 2% and 10% by making use of the probability theory. It should be explained that such a calculation assumes food surplus and shortage as probability incidents, including the impact of weather on food output, the impact of price fluctuations, the impact of international market on domestic food demand and supply, the impact of cross-substitution (such as, the decrease of meat supply will result in the increase of food demand, etc.), and other factors. This calculation cannot explain sudden accidents, such as, the impacts of wars and volcanic eruptions on food production, inadequate food supply caused by civil wars and large-scale invasions, etc.

The specific calculation method is as follows:

Step 1: Normal Distribution Test. We will calculate the relative number of the grain gap from 1978 to 2006 based on the three sets of data in Method I,

Method II and Method III so as to test if the ratio of grain gap to grain output and the ratio of grain gap to grain consumption conform to normal distribution. Intuitive means to judge if data conforms to normal distribution include using frequency distribution histograms to test if data shows skew distribution, QQ and PP plots, and Shapiro-Wilk W test which is a more accurate econometric method.

Step 2: Calculation of the amount of grain gap. If the tested data conforms to normal distribution, the following figure is the graph of the normal distribution. The horizontal axis x shows the ratio of grain gap to grain output. The value of x above zero represents grain surplus while the value of x below zero represents grain gap.

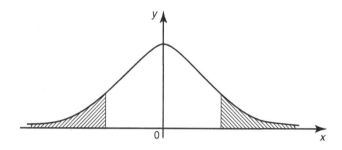

Let us take the calculation of the ratio of grain gap to grain out in the case of the occurrence of grain gap at the probability of once in 100 years (or 1%) as an example. The algorithm is as follows: we firstly calculate mean μ and variance σ^2.

By using the standard normal distribution table, we can ascertain the value of x when the probability is 0.01. Then we can calculate the value of x' in accordance with the formula of $x = \frac{x'-\mu}{\sigma}$, that is, the ratio of grain gap to grain output when grain shortage, which happens once in 100 years, occurs.

In the same way, we can calculate the ratio of grain gap to grain output or grain consumption when grain shortage, which happens once in 50 years (at the probability of 2%), occurs and the ratio of grain gap to grain output or grain consumption when grain shortage, which happens once in 10 years (at the probability of 10%), occurs.

4. Calculation Results

As the accuracy of Method I is the highest, the main basis of the following calculation also comes from this set of data.

Step 1: Normal Distribution Test

Based on the data of Method I (Data I), we have the following descriptive statistics of the relative number of grain gap:

Summarize shortage rate Variable	Obs	Mean	Std. Dev.	Min	Max
Ratio of Grain Gap to Grain Output	29	0.0170724	0.0570726	−0.129	0.1201
Ratio of Grain Gap to Grain Consumption	29	0.0206403	0.0585566	−0.1142416	0.1364359

Generally speaking, there are three intuitive ways to judge if data conforms to normal distribution: frequency distribution histograms, PP plot and QQ plot. Next, we will make an analysis from the following three means.

- Judging if data shows skew distribution from frequency distribution histograms.

The histogram and the normal distribution curve of the relative number of grain gap are as follows:

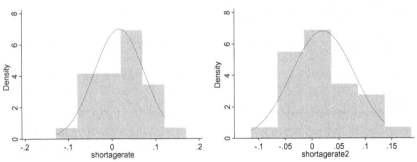

Frequency distribution histograms: the ratio of grain gap to grain output (left), the ratio of grain gap to grain consumption (right). (Data I).

- QQ plot and PP plot

If the scattered points plot along the fixed straight line, we can conclude that the data roughly conforms to normal distribution.

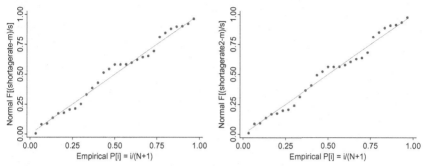

PP Plot: the ratio of grain gap to grain output (left), the ratio of grain gap to grain consumption (right). (Data I).

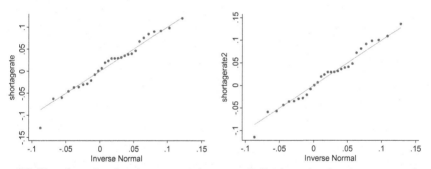

QQ Plot: the ratio of grain gap to grain output (left), the ratio of grain gap to grain consumption (right). (Data I).

From the above frequency distribution histograms, PP plots and QQ plots, we can see that the relative number of grain gap in Data I soundly conforms to normal distribution. In order to verify the goodness of fit of the normal distribution for data in terms of statistics, we conduct the following Shapiro-Wilk test.

- Shapiro–Wilk Test

Introduced in 1965, this test is also called W test and especially suitable for testing the normality of data with a sample size of less than 50. If the

p-value in the normal distribution test of data is smaller than 0.05, we can conclude that the data shows skewness, or that the data does not conform to normal distribution. The W test results are as follows:

Swilk shortage

Shapiro–Wilk W test for normal data					
Variable	Obs	W	V	z	Prob > z
Ratio of Grain Gap to Grain Output	29	0.97500	0.775	−0.526	0.70069
Ratio of Grain Gap to Grain Consumption	29	0.98170	0.567	−1.170	0.87892

The W test shows that the p-value of the ratio of grain gap to grain output is 0.70069 and the p-value of the ratio of grain gap to grain consumption is 0.87892, far higher than 0.05. Therefore, from the above QQ plot, PP plot and the W test, we can see that the relative number of grain gap conforms to normal distribution. Next, we need to calculate the grain gap when food shortage happens once in 100 years, 50 years and 10 years or at the probability of 1%, 2% and 10% occurs.

Step 2: Calculation Results
In accordance with the abovementioned algorithm, we get the following result: the mean μ of the ratio of grain gap to grain output of 29 years is 0.0170724; the variance σ^2 of the ratio of grain gap to grain output is 0.05707262; the mean μ of the ratio of grain gap to grain consumption of 29 years is 0.0206403; and the variance σ^2 of the ratio of grain gap to grain output is 0.05855662.

By using the standard normal distribution table, we find out that the value of x when the probability is 0.01 is −2.33, the value of x when the probability is 0.02 is −2.05 and the value of x when the probability is 0.1 is −1.28. In accordance with the formula $x = \frac{x'-\mu}{\sigma}$, we get $x' = x\sigma + \mu$.

In accordance with the above calculation, we conclude that:

- In the case of once in 100 years, the ratio of food gap to food output is $-2.33 \times 0.0570726 + 0.0170724 = -0.11591$, which can be explained as follows: The ratio of food gap to food output would be 11.6% if the food shortage that happens only once in 100 years occurs. The ratio of food gap to food consumption is $-2.33 \times 0.0585566 + 0.0206403 = -0.1158$, which can be explained as follows: the ratio of food gap to food consumption would be 11.6% if the food shortage that happens only once in 100 years occurs.

- In the case of once in 50 years, the ratio of food gap to food output is $-2.05 \times 0.0570726 + 0.0170724 = -0.09993$, which can be explained as follows: the ratio of food gap to food output would be 9.99% if the food shortage that happens once in 50 years occurs. The ratio of food gap to food consumption is $-2.05 \times 0.0585566 + 0.0206403 = -0.0994$, which can be explained as follows: the ratio of food gap to food consumption would be 9.94% if the food shortage that happens once in fifty years occurs.

- In the case of once in 10 years, the ratio of food gap to food output is $-1.28 \times 0.0570726 + 0.0170724 = -0.05598$, which can be explained as follows: the ratio of food gap to food output would be 5.60% if the food shortage that happens once in 50 years occurs. The ratio of food gap to food consumption is $-1.28 \times 0.0585566 + 0.0206403 = -0.0543$, which can be explained as follows: the ratio of food gap to food consumption would be 5.43% if the food shortage that happens once in 10 years occurs.

5. Conclusion

In summary, judging from the food gap reflected by the three sets of data, Method I is more reliable. This is because in Method II, the data of food stock cannot be verified whereas many scholars have made use of Method I to calculate China's food demand due to the feasibility of its deduction method although there's no precise description in statistics with regard to data of food demand. In addition, although data of the Department of Agriculture of the U.S. is not the total amount of food in the strict sense, it was also calculated in Method I and showed a changing trend of food gap similar to that reflected by domestic data which was also calculated in Method I. Therefore, this set of data is a powerful complement to Method I.

Now, let us look at the method used in this report. Data calculated in Method I, including the ratio of food gap to food output and the ratio of food gap to food consumption conforms to normal distribution. Therefore, we believe that Method I is optimal both in terms of the direct statistical result and in terms of calculation methods of food gap in various conditions in the stricter sense.

We will make an analysis of China's food gap based on these judgments. It can be seen that, at present, China's food output basically equals to its food consumption at 0.5 billion tons. For instance, China's food output stood at 10.52 million tons in 2006, accounting for 2.1% of China's food output of that year. In the past 30 years, the ratio of food output to food output stood around 5% in most years, with the highest being 12.9% in 2003; the ratio of food gap to food consumption reached the peak at 11.4%, also in 2003. Besides, according to the calculation in this report, when food shortage occurs at the probability of once in 100 years, food gap will account for 11.6% of food output and 11.6% of food consumption that year.

At present, China has abundant food reserve. FAO proposed that the minimum safe level of world stocks for all cereals would need to be within the range of 17%–18% of the global cereals consumption. China has long exceeded this minimum level. If food shortage occurs in China, direct food stock could be used to solve the problem to a large extent. Even excluding the probability of settling the problem with food stock, we can see that China has ample foreign exchange reserve to cope with the problem after we calculate the foreign exchange required to purchase grain and make up the gap at the high grain price at present.

Chapter 10

Domestic Solutions to China's Food Shortage

1. Introduction

With the largest population in the world, China's per capita cultivated land area is less than half of the global average. This long-term reality determines that food security will remain one of the focuses of Chinese government and society.

Since its reform and opening-up, China's aggregate food output has risen by 64.5% from 0.305 billion tons in 1978 to 0.502 billion tons in 2007 with an annual growth rate of 1.74% which was higher than the growth rate of population during the same period. Due to the incentive of price recovery and the encouragement of fiscal supporting policies by the government, China's food output witnessed a rapid growth of 16.4% from 2004 to 2007. In terms of different grain varieties, paddy output and wheat output witnessed small increases whereas corn output and soybean output underwent substantial growth as feed grains (see Figs. 10.1 and 10.2). This is mainly because of different demand for different varieties of crops.

Under the context of the substantial increase of per capita food output, the real grain price showed a declining trend. This reflected that China's food situation had changed from supply restraint to demand restraint from one aspect and also partially explained the slow growth of paddy output and wheat output as mentioned above.

Meanwhile, we should notice some problems in the field of food of China. After maintaining a steady growth of nearly 20 years, China's food output witnessed a drastic drop from 2000 to 2003, descending by 15% from 508.39 million tons to 430.7 million tons. China's food price went up by 8% from 2006 to 2007 and is likely to continue to rise in the future.

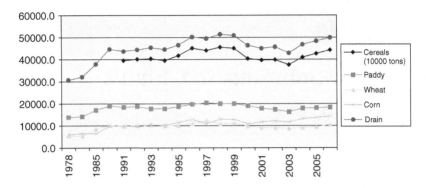

Figure 10.1. China's Aggregate Grain Output and Its Decomposition, 1978–2006. *Sources: China's Agricultural Statistics 2005; data of 2006 comes from China Grain Development Report 2007.*

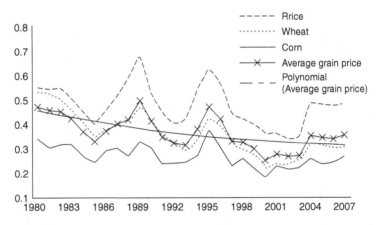

Figure 10.2. Real Grain Price in Domestic Market of China (RMB/jin). *Sources: Lu Feng (2007).*

This will affect the social stability of China in which millions of people are suffering from poverty.

Under the background of soaring international energy prices and the development of renewable energy, countries throughout the world are vigorously advocating the production of biofuels. What will be the trend of world food prices? How should China balance the relationship between the development of bioethanol and food security? These new problems add new uncertainty to China's food problems which are already very complicated.

Under this background, many officials and scholars predict that the relationship between China's food supply and demand will remain tight for quite a long time even with the possibility of food crisis. The opinion also affected the government to a certain extent. The anxiety over food problems caused the government to take an excessively cautious attitude towards food problems. For example, under the background of rapid urbanization and industrialization in China, Chinese government persists in regulating the amount of cultivated land and sticks firmly to the red line of 1.8 billion mu.

This chapter aims to answer the question of how China could solve its food security problems by domestic means. The structure of this chapter is as follows: Section 2 is about China's food demand situation and the prediction; Section 3 is about the institutional and scientific factors in food production; Section 4 is about the substitution of different factors of food production; Section 5 is about the analysis of the role of food stock in smoothing fluctuations of food supply; and Sec. 6 is the conclusion.

2. China's Food Demand Situation and the Prediction

China's grain demand includes ration demand, feed demand, industrial demand and seed demand. Due to lack of systematic data on food consumption, we estimated the proportions of ration consumption, feed demand and industrial demand at 6:3:1 on the basis of the collected data from different sources. Seed consumption is omitted as its proportion in the aggregate food consumption has been around 2%. In terms of trends, ration consumption remained rather stable while feed consumption and industrial consumption witnessed rapid growth despite their small initial amounts. In terms of grain varieties, in 2006, the total consumption of paddy stood at 180 million tons, among which 147 million tons or 81.7% were used for food consumption; domestic demand for wheat stood at 99.52 million tons, among which 86.5 million tons or 86.9% were used for flour production, 2.25 million tons were used for industrial consumption and 4 million tons were used for feed consumption; domestic demand for corn sat at 141 million tons, among which 35.5 million tons or 25.2% were used for industrial consumption and 94 million tons or 66.7% were used for feed consumption; domestic demand for soybean stood at 44.03 million tons, among which 34.7 million tons were used for oil extraction and 8.06 million tons were used for food and industrial consumption.

It can be seen that corn constitutes a major part of feed grain and industrial grain which were largely affected by prices while paddy and wheat serve as major rations and constitute a rather small proportion of feed grain and industrial grain. Therefore, the rapid growth of feed demand and industrial demand for food do not impose much pressure on ration demand for grain.

2.1. *Grain demand for different uses*

2.1.1. *Ration*

From the Fig. 10.3 it can be seen that the per capita grain consumption showed a declining trend in the past 20 years and descended at an accelerated rate in the past 10 years.

Now, we will analyze various factors affecting ration demand and predict the future ration demand.

✧ Demand Structure

When we analyze the food demand of a transitional country, we should also take into account the structural changes of food consumption brought by changes in consumer preference.

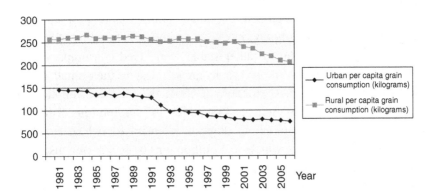

Figure 10.3. The Per Capita Grain Consumption of China, 1980–2006.
Sources: China Rural Statistical Yearbook 2007.

Jikun Huang and Howarth Pouis (2001) listed five structural factors affecting food demand, including:

- There may be a wider choice of foods available in urban markets.
- Urban residents are more likely to be exposed to the rich variety of dietary patterns of foreign cultures.
- Urban life styles may place a premium on foods that require less time to prepare, for example, if employment opportunities for women improve and the opportunity cost of their time increases.
- Urban occupations tend to be more sedentary than rural ones.
- Farmers tend to eat more of the food grown by themselves due to price differentials.

Due to the urban-rural dual structure, the voucher-based supply of food was not abolished until the early 1990s in China. The artificially slowed urbanization process also postpones the declining trend of ration consumption brought by urbanization. In the past 10 years, with economic development and the accelerated urbanization process, we predict that the overall ration consumption will show a decreasing trend under the background of steady population growth.

This point could be further verified by the above data. The grain consumption of Chinese urban and rural residents has been declining year by year whereas their consumption of meat, poultry, aquatic products, milk experienced a substantial increase. Therefore, we believe that the future demand for ration crops (paddy and wheat) faces smaller pressure than the future demand for feed crops (corn).

From Figs. 10.4–10.6, we can see that grain consumption showed declining trends in two aspects. First, the proportion of food consumption in the daily consumption of residents has been declining. Second, the proportion of grain consumption in the overall food consumption has been decreasing.

✧ Income Effect

In Lu Feng (2007), the estimations of the income elasticity of rice, wheat, corn and soybean showed that the income elasticity of rice and wheat in China has declined, with time, to below zero. We predict that this trend will continue for a period of time until it stabilizes at a certain level.

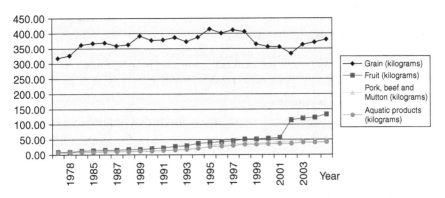

Figure 10.4.   The Per Capita Agricultural Product Output of China.
Sources: *China Rural Statistical Yearbook 2007.*

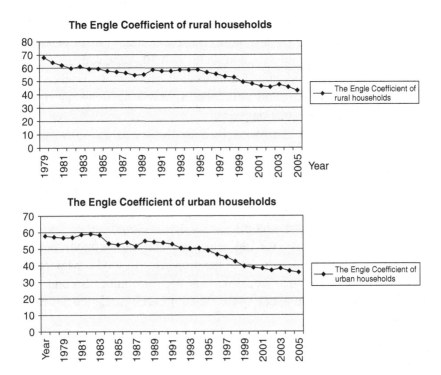

Figure 10.5.   The Engle Coefficients of Rural and Urban Households.
Sources: *Database of the National Bureau of Statistics of China.*

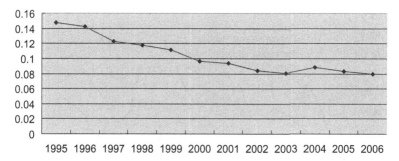

Figure 10.6. The Ratio of Rural Residents' Grain Consumption in their Food Consumption, 1995–2006.
Sources: Database of the National Bureau of Statistics.

At the present stage, grain could be seen as Giffen goods, the consumption of which descends while the income of residents ascends (see Fig. 10.7).

The neglect of structural changes will lead to overestimation of income and income elasticities. It should also be noted that the change in grain assumption is not so significant in some transitional countries. For example, the per capita rice consumption of South Koreans declined from 140 kilograms to 126 kilograms from 1973 to 1992 while the per capita GDP grew by over 30 times during the same period.

✦ Population Growth

Lu Feng (2007) summarized the analysis of the trend of China's population in the future.

Based on statistics in Table 10.1, Lu Feng (2007) deduced that the ceiling of China's future annual grain demand will be around 0.55 to 0.58 billion tons and unlikely to exceed 0.6 billion tons.

To summarize the above analysis, based on the negative effect of structural change of food consumption, the negative growth of food grain consumption caused by income increase and the growth trend of population, we predict that China's future ration consumption will not be much higher than the present level.

2.1.2. *Feed grain*

The growth rate of China's feed grain is lower than that of animal products for the following three major reasons. First, feed conversion rate has

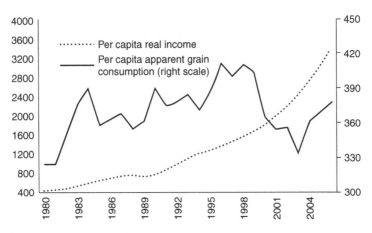

Figure 10.7. The Per Capita Real Income and Per Capita Apparent Grain Consumption. *Sources and note: Lu Feng (2007).* Per Capita Apparent Grain Consumption = (Grain Output + Net Grain Import)/Population. Please refer to page 107 of China Grain Development Report 2007 for data of grain output and page 131 and 133 of China Grain Development Report 2007 for data of grain import and export. Data of population comes from China Labor Statistical Yearbook 2006. The per capita real income was computed by deflating the per capita nominal income from Table 3-1 of China Statistical Yearbook 2007 by consumer price index (CPI) (1978 as the base year). The CPI from 1978 to 2005 is from the Yearbook of the People's Republic of China 2006, and the CPI of 2006 is from China Statistical Yearbook 2007.

improved. Second, the structural change of consumption of animal products has undergone a change, namely, people have shifted their consumption of animal products from pork and chicken to beef and mutton which use up less feed grain. Feed grain consumed by pigs and chicken is four to five times that consumed by cattle and sheep for production of per 50 kilograms of animal products. Third, the by-products of industrial grain will increase feed supply.

2.1.3. *Industrial grain*

Industrial grain in 2006 included 20 million tons of paddy, 2.25 million tons of wheat, 35.5 million tons of corn and 42.76 million tons of soybean (including oil extraction, food consumption and industrial use).

The major raw materials of biofuels include corn, cassava, sweet sorghum and waste oil. The medium and long-term development plan for renewable energy clearly stated that China's production capacity of fuel

Table 10.1. Analysis of the Trend of China's Population in the Future.

Peak population (0.1 billion)	Peak years	Source	Notes
14.48	2030	Huang Baofeng (1999)	Median prediction by assuming TFR stood between 1.5 and 2.0.
13.95	2050	UN (2004)	The year 2030 will be a turning point. The peak population will be 1.4 billion.
15.58	2030	UN (2007)	
13.90	2050	Men Kepei *et al.* (2004)	Prediction of net population increase by the grey dynamic model.
13.73–14.14	2030	Jiang Hui (2005)	
14.39	2030	Cai Fang (2006)	Research results of China Population and Development Research Center
14.69	2029	Guo Zhigang (2006)	
14.81	2040	Zeng Yi (2006)	Median prediction
14.40	2030	Chen Wei (2006)	
15亿	2033	Research Report on National Population Development Strategy	In the figure the peak population stood below 1.5 billion whereas the text description is "around 1.5 billion".

ethanol would reach 10 million tons and its production capacity of bio-diesel would reach 2 million tons by 2020 so as to substitute 10 million tons of oil products.

Prior to 2020, the production capacity of grain-based fuel ethanol in China will be controlled below 1.5 million tons, basically equal to the production capacity of four designated enterprises of production of fuel ethanol in China. The state will not examine and approve any new or newly expanded projects of grain-based fuel ethanol any more and the focus of fiscal subsidies will also transfer to projects of non-grain raw materials. The newly added production capacity of 4.2 million tons of fuel ethanol by 2010 will also use cassava, sweet sorghum, straw and other non-grain crops as raw materials.

However, grain producing provinces hope to form grain industrial chains by completing the deep processing of grain within these provinces rather than to transport raw grain out of these provinces. For instance, in the Opinions on Promoting the Development of Grain Processing Industry (Hei Zheng Fa (2006) No. 2) issued by Heilongjiang Province, it was clearly stated "try the best to change the situation of transporting large amounts of raw grain out of Heilongjiang province within two to three years". The total production capacity of corn deep processing projects in Jilin has reached 9.45 million tons by now and is forecast to reach 18.45 million tons by 2010 so as to use up the total corn output of Jilin province.

In view of this problem, National Development and Reform Commission issued two circulars regarding the development of bioenergy, namely, the Urgent Circular on Strengthening the Construction Management of Corn Processing Projects (Fa Gai Gong Ye (2006) No. 2781) and the Circular of National Development and Reform Commission and the Ministry of Finance on Strengthening the Construction Management of Bio-fuel Ethanol and Promoting the Healthy Development of Industries (Fa Gai Gong Ye (2006) No. 2842).

However, according to the deduction, the production of bioenergy will be profitable as long as the oil price exceeds $40 per barrel. Therefore, it's very difficult to solve the problem of the competition for grain between bioenergy production and human consumption merely through administrative means without making use of price mechanism.

In accordance with Datagro, the largest sugar prediction organization in Brazil, the competitive demand for corn between the food industry and the ethanol industry in China will lead to the decline of corn export by China and finally turn China into a corn importing country.

2.1.4. *Seed grain*

The total annual seed grain stabilized at around 10 million tons in recent years.

2.2. *Prediction of grain demand*

In summary, we can see that serious problems do not exist in China's ration demand, the most import aspect of food security. Industrial grain and feed grain have little influence on food security and are greatly affected by their own prices and the prices of their substitutions.

Huang Jikun (2004) had forecast China's agricultural product self-sufficient rates under plans of different agricultural inputs by using the China's Agricultural Policy Simulation and Projection Model (CAPSiM, see appendix for its introduction) developed by the Center for Chinese Agricultural Policy of CAS.

The base scheme assumes that China would enter the WTO and fulfill its commitments toward WTO and that the new Doha Round negotiations would achieve the intended goals. In the scheme of high investment in scientific research and irrigation, it is assumed that China would gradually increase its investment in agricultural science and technology and irrigation, with the annual growth rate of agricultural scientific research investment rising from 5% in the base scheme to 7% and the annual growth rate of irrigation investment going up from 4% in the base scheme to 6% in the following 20 years. In the scheme of low investment in scientific research and irrigation, it is assumed that the annual growth rates of agricultural scientific research investment and irrigation investment would reach 3.5% and 2.5% respectively in the following 20 years (see Tables 10.2 and 10.3).

Table 10.2. The Self-Sufficient Rates of China's Major Agricultural Products in the Base Scheme (%), 2001–2020.

	2001	2010	2020
Rice	101	102	106
Wheat	99	90	96
Corn	105	80	72
Soybean	61	64	60
Oil crops	84	69	66
Sugar	96	83	76
Vegetable	101	103	104
Fruit	100	105	107
Cotton	100	94	86
Pork	101	110	105
Beef	100	98	96
Mutton	99	96	95
Poultry	99	106	105
Milk and its products	97	83	79
Aquatic products	105	110	107

Table 10.3. Outputs and Self-Sufficient Rates of China's Major Agricultural Products in 2020 under Conditions of Different Investment in Agricultural Product Scientific Research and Irrigation.

	Percentage of output changes in comparison with the base scheme (%)		Self-sufficiency rate (%)	
	High investment in scientific research and irrigation	Low investment in scientific research and irrigation	High investment in scientific research and irrigation	Low investment in scientific research and irrigation
Grain	2.86	−3.27	92	86
Cereals	2.75	−3.14	91	86
Paddy	1.78	−2.05	108	104
Wheat	2.8	−3.18	98	93
Corn	3.72	−4.27	75	68
Soybean	4.83	−5.46	65	55
Oil crops	6.1	−6.87	70	61
Sugar	4.36	−5.04	80	72
Vegetable	5.54	−6.27	109	99
Fruit	8.33	−9.16	115	98
Cotton	4.78	−5.54	91	82

This predication also confirms that problems will not occur to China's food security in the future as long as appropriate investment is ensured. At the same time, it should be avoided that excessively high price is paid for the overemphasis of food security; industrial grain and feed grain absolutely could maintain low self-sufficient rates.

3. Institutional and Scientific Factors in Grain Production

The changes in grain output from 1979 and 2003 are reviewed in the following passages with the emphasis on three import factors affecting grain production, including, institution, technology and input. This part mainly referred to Chen Weiping and Zheng Fengtian (2006).

1979–1984

At this stage, the TFP of grain in China experienced the fasted growth from 1953 to 2003 mainly because of institutional reforms, including the replacement of the production brigade with rural household responsibility system and the opening of the market of agricultural products and factors of production. According to J. Y. Lin (1992), rural household responsibility system is the major contributing factor. During this period, paddy output went up at an annual average growth rate of 4.49%, paddy TFP by 6.69% and the total input in paddy production by −2.06%; wheat output rose at an annual average growth rate of 8.5%, wheat TFP by 10.7% and the total input in wheat production by −2% (according to Jin *et al.* (2002), this is because the labor input witnessed a decrease at this stage); corn output increased at an annual growth rate of 4.63%, corn TFP by 8.37% and the total input in corn production by −3.45%.

1985–1991

In 1985, the unified procurement and distribution of grain was abolished and contract procurement was implemented; grain outside the contract procurement was allowed to be freely traded in the market. The excessively high contract procurement price dampened farmers' enthusiasm for grain production and led to the slow increase in grain TFP in China from 1985 to 1991. During this period, paddy output went up at an annual average growth rate of 0.44%, paddy TFP by −0.11% and the total input in paddy production by 0.55%; wheat output rose at an annual average growth rate of 1.27%, wheat TFP by −0.14% and the total input in wheat production by 1.41%; corn output increased at an annual growth rate of 4.33%, corn TFP by 1.6% and the total input in corn production by 2.69%.

1992–1996

The TFP of China's major grain resumed growth mainly because of a series of agricultural reform measures in the 1990s. Major reform measures during this period include: the average price of major grain crops increased by 18% in 1989; special attention was paid to ensure the supply of fuels, electric power and major raw materials for production, such as, fertilizer, pesticide and agricultural plastic film to farmers; the system of unified procurement and distribution of grain was abolished, while the policy of "guaranteeing grain procurement target while liberalizing the grain purchase prices" and the system of provincial governors assuming responsibility for the rice bag were issued successively. Apart from the above-mentioned reforms, the adoption of new grain varieties by farmers was also an important factor.

For example, the investigation of Jin *et al.* (2002) showed that farmers substituted an average of 20% to 25% of grain varieties annually. This is also one of the major reasons of the recovery of grain TFP during this period.

During this period, paddy output accelerated at an annual average growth rate of 1.2%, paddy TFP by 2.42% and the total input in paddy production by −1.19%; wheat output rose at an annual average growth rate of 1.27%, wheat TFP by −0.14% and the total input in wheat production by 1.41%; corn output increased at an annual growth rate of 4.33%, corn TFP by 1.6% and the total input in corn production by 2.69%.

1997–2003

Certain changes had affected the market environment of grain production at this stage. The previous resource constraint for grain production was replaced by the double constraints of resource and market. Therefore, fiscal input in agricultural scientific research and agricultural infrastructure was enlarged.

During this period, paddy output went up at an annual average growth rate of −2.74%, paddy TFP by 1.71% and the total input in paddy production by −4.34%; wheat output rose at an annual average growth rate of 1.2%, wheat TFP by 2.42% and the total input in wheat production by −1.91%; corn output increased at an annual growth rate of −1.36%, corn TFP by 0.83% and the total input in corn production by −2.17%.

4. The Substitution of Different Factors of Grain Production

4.1. *Problems*

In the general sense, grain production requires inputs of intermediate products (including seeds, seedlings, chemical fertilizer, organic fertilizer, agricultural plastic film, pesticide, animal power, machinery, equipments and other material inputs), land and labor. Obviously, when the cultivated land area remains unchanged, improvements in the multiple-cropping index could realize the increase in the grain sown area, thus leading to an increase in the land input.

Next, we will try to assess the quantum of intermediate products and labor that should be inputted to compensate for the reduction in the grain

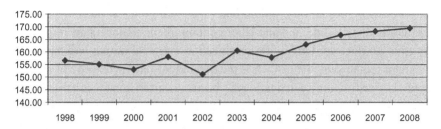

Figure 10.8. The Multiple-Cropping Index, 1998–2008.
Sources: *The China CropWatch System.*

sown area of 1%, 5% and 10% and the occurrence of crop failure at the probability of once in 10 years, once in 50 years and once in 100 years.

4.2. Method

It is necessary to lay the hypothesis for our further discussion:

✧ Hypothesis 1: We use the Cobb-Douglas production function to analyze China's grain output.

$$Y_t = A_t (K_t)^{\alpha_t} (L_t)^{\beta_t} (T_t)^{\gamma_t} \tag{1}$$

Where Y_t is the total output of period t; A_t is the total factor productivity (TFP) of period t; K_t, L_t and T_t are the intermediate input, labor input and land input of period t respectively.

✧ Hypothesis 2: We hypothesize that A_t (TFP), r (the price of intermediate products) and w (the price of labor) all remained unchanged after we increase the inputs of intermediate products and labor.
✧ Hypothesis 3: We hypothesize that all micro subjects have made the optimal decisions regarding the observed prices, or, the output elasticities of all factors equal to the ratios of their prices.
✧ Hypothesis 4: China's grain output is considered as having constant returns to scale under current technological level.

We made this hypothesis because of the following factors: (1) As China is a large grain producing country and the improvement in productivity brought by further deepened division of grain production could be omitted, the newly added input could be used to engage in the same production activities as before and the output and input will go up at the same rates; (2) Intermediate input, land and labor contain all the major inputs required for grain production. Otherwise, diminishing marginal output will occur if one input remains unchanged while the other inputs are increased. Since the input of infrastructure i.e. irrigation has been converted into the intermediate input, we can safely assume that no major grain production input factors have been left out in the following discussion.

We suppose the production function has constant returns to scale. That is, $\alpha_t + \beta_t + \gamma_t = 1$, where α_t, β_t, γ_t represent the shares of intermediate input, labor and land in the total input respectively and equal to the output elasticity of each factor.

Given hypothesis 1 to 3, we derive that the optimal ratio of intermediate input to labor is:

$$K = \frac{w}{r} \cdot \frac{\alpha}{\beta} \cdot L \qquad (2)$$

In order to calculate the input increase required by output increase, we divide the production function of Y_1 by the production function of Y_2:

$$\frac{Y_1}{Y_2} = \left(\frac{K_1}{K_2}\right)^{\alpha} \cdot \left(\frac{L_1}{L_2}\right)^{\beta} \cdot \left(\frac{T_1}{T_2}\right)^{\gamma} \qquad (3)$$

We insert equation (10.2) into equation (3)

$$\frac{Y_1}{Y_2} = \left(\frac{L_1}{L_2}\right)^{\alpha+\beta} \cdot \left(\frac{T_1}{T_2}\right)^{\gamma} \qquad (4)$$

Next, we need to calculate α, β and γ in equation (4)

4.3. *Calculation of output elasticities*

We mainly adopted two data sources in this chapter, including the survey data of agricultural product costs from the National Development and Reform Commission (NDRC) and the data of China Agriculture Yearbook over the years. The input of intermediate products is the sum of (1) seed cost; (2) chemical fertilizer cost; (3) organic fertilizer cost; (4) pesticide cost; (5) agricultural plastic film cost; (6) leasing and operation costs (machinery operation cost, irrigation and drainage costs (including water charge), animal power cost); (7) fuel and power costs; (8) technical

Table 10.4.   Output Elasticities of the Factors, 1990–2006.

Year	Intermediate products	Labor	Land
1990	0.60	0.36	0.07
1991	0.58	0.38	0.07
1992	0.57	0.39	0.08
1993	0.58	0.37	0.08
1994	0.62	0.33	0.09
1995	0.58	0.37	0.09
1996	0.54	0.41	0.09
1997	0.55	0.41	0.08
1998	0.54	0.37	0.14
1999	0.55	0.37	0.14
2000	0.57	0.35	0.13
2001	0.54	0.39	0.13
2002	0.54	0.37	0.15
2003	0.54	0.37	0.15
2004	0.53	0.38	0.14
2005	0.50	0.36	0.15
2006	0.50	0.34	0.15

service cost; (9) instrument and material costs; (10) maintenance and repair costs; (11) other direct costs; (12) depreciation of fix assets; (13) tax; (14) insurance fee, (15) administration fee; (16) financial expense; (17) sales expense. The inputs of labor and land are converted from wage and land rent.

For the sake of the comparability and completeness of data, we adopted data from 1991 to 2006 and excluded data from 1978, 1985, 1998 and 1990.

We can directly obtain the output elasticities of the factors after the calculation of the total expense of grain production and the proportions of each factor (see Table 10.4).

4.4. *Input increase required to make up for the grain output decrease under conditions of different sown area reductions*

We will insert the average of elasticities of 2004, 2005 and 2006 into equation (4) and calculate the input increases of intermediate products and labor required to make up for the insufficient grain output caused

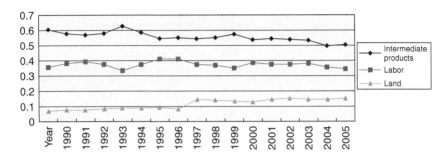

Figure 10.9. Output Elasticities of the Factors, 1991–2006.

Sources: Monitoring data of costs released by the National Development and Reform Commission.

Table 10.5. Input Increase Required to Make up for Grain Output Reduction When Grain Sown Area Remains Unchanged.

Gap of grain supply and demand (%)	Required input increase of intermediate products (%)	Required input increase of labor (%)
−11.6	13.43	13.43
−9.99	11.56	11.56
−5.6	6.46	6.46
−11.6	13.43	13.43
−9.94	11.50	11.50
−5.43	6.26	6.26

Table 10.6. Input Increase Required to Make up for Grain Output Reduction When Grain Sown Area Decreases by 1%.

Gap of grain supply and demand (%)	Required input increase of intermediate products (%)	Required input increase of labor (%)
−11.6	13.63	13.63
−9.99	11.75	11.75
−5.6	6.64	6.64
−11.6	13.63	13.63
−9.94	11.69	11.69
−5.43	6.44	6.44

Table 10.7. Input Increase Required to Make up for Grain Output Reduction When Grain Sown Area Decreases by 5%.

Gap of grain supply and demand (%)	Required input increase of intermediate products (%)	Required input increase of labor (%)
−11.6	14.43	14.43
−9.99	12.54	12.54
−5.6	7.39	7.39
−11.6	14.43	14.43
−9.94	12.48	12.48
−5.43	7.19	7.19

Table 10.8. Input Increase Required to Make up for Grain Output Reduction When Grain Sown Area Decreases by 10%.

Gap of grain supply and demand (%)	Required input increase of intermediate products (%)	Required input increase of labor (%)
−11.6	15.49	15.49
−9.99	13.58	13.58
−5.6	8.38	8.38
−11.6	15.49	15.49
−9.94	13.52	13.52
−5.43	8.18	8.18

by the occurrence of crop failures at the probability of once in 10 years, once in 50 years and once in 100 years and under the conditions of grain sown area remaining unchanged and reducing by 1%, 5% and 10%. The calculation results are shown in the following tables from Table 10.5 to 10.8.

5. An Analysis of the Role of Grain Stock in Smoothing Grain Supply Fluctuations

It can be seen from the above passages that China's long-term grain demand will not increase significantly above the current levels and that China has a large potential of grain output increase as long as we ensure appropriate policies and inputs. That is to say, in terms of long-term trend, China does not have serious problems in food security. However, in the short term, crop failures might occur in particular years due to factors such as weather; the rapid decline of grain prices (in some particular years) might dampen farmers' enthusiasm for production and cause grain production to descend in the following years. The fourth chapter of this book has discussed how to solve this problem through international trade. In the following passage we will discuss how to adjust and smooth grain output fluctuations between different years by making use of domestic food stock.

Theoretically speaking, grain stock helps to reduce fluctuations in grain output, to further stabilize grain prices and to reduce market uncertainties. The reality, however, is that the state has kept the situation and relevant statistics of China's grain stock as secret information which was never disclosed to the public and that the handling and operation methods of grain stock also lack in transparency. This enlarges the uncertainly of grain information. Apart from the problems mentioned above, two other problems exist in China's grain stock management. First, the distribution of grain stock is unreasonable. The grain stock of 14 major grain producing regions accounted for over 80% of the national grain stock in 2002. Second, the storage period of grain exceeds the normal storage time limits. China's grain stock beyond the normal storage time limits made up 20% of the total grain stock in 2002.

According to various estimations, the ceiling of China's annual grain stock capacity was around 0.3 million tons, accounting for over 50% of China's annual grain consumption. The regular annual grain stock stood around 0.2 million tons. These two figures are far higher than the minimum grain stock safety range of 17% to 18% suggested by FAO. In accordance with the calculation of Ma Jiujie and Zhang Chuanzong (2002), the ceiling of China's grain stock of 0.3 billion tons could completely smooth out China's grain output fluctuations between different years.

Nevertheless, it should also be noted that China's high grain stock was maintained on the basis of current grain production conditions. If China decides to lower its grain self-sufficiency rate and relies more on grain import in the future, further research would be required to study if it is indeed feasible to maintain such a high grain stock through international trade and what difficulties will China encounter.

6. Conclusion

From the above analysis, it can be seen that the increase in China's grain demand is limited in the longterm. China's grain supply has a large potential through more investment in research, development and infrastructure and that China's current grain reserve system definitely could cope with extreme grain production fluctuations in the short-term. Therefore, China's food security problems could be totally solved through domestic means. Meanwhile, from the cost-benefit perspective and under the premise that China's food security should not be affected, land policies regarding rural land could be relaxed to a reasonable extent.

Chapter 11

International Solutions to China's Food Shortage

1. Background

The contradiction between global food supply and demand has become increasingly conspicuous since 2006, with international food prices rising by 9.8% year-on-year. World cereal prices continued to go up substantially in 2007, with wheat price increasing by 112%, corn price by 47% and soybean by 75%. China's food prices ascended along with world food prices, with the domestic cereal price going up by 3.4% and 4.3% quarter on quarter in the third quarter and fourth quarter of 2006 respectively. China's cereal prices went up by a larger scale in 2007, with a quarter-on-quarter increase of 8.0%, 8.0%, 7.5% and 10.6% in four quarters respectively (Analysis and forecast on China's Rural Economy, 2008). The rising prices globally enabled the topic of "Who Will Feed China" raised by Mr. Brown to prevail again. More worldwide attention was attracted to China's food security problem. If a food gap occurs in China and if domestic means are inadequate to settle this gap, it will be prudent to consider reducing this gap through the international market.

1.1. *The correlation between China's food trade and global food prices*

The global food prices have experienced a rapid rise since 2006, which has been attributed by many people in the international community to "China factor". This saying, however, completely contradicts the facts. China is an important stabilizing factor, and not the driving force of the soaring world food prices (Ding Shengjun, 2008). In sharp contrast to the decline of global food output and reserve, China's food output went up continuously, leading to balanced food supply and demand, increased food reserves and ample market supply of food. China's food price merely experienced a mild and structural rise. From 2004 to 2007, China's aggregate food output continued to increase during these consecutive years, from 469.469 million tons in 2004 to 484.022 million tons in 2005, 497.499 million tons in 2006 and 501.50 million tons in 2007. China's food reserve has reached a rather high level, accounting for over 35% of China's annual food consumption and far higher than the minimum safety range of 17%–18% proposed by FAO. Meanwhile, China has maintained a high food self-sufficiency rate which always stood above 95%. With a decreased import volume of cereals (excluding soybean) and an increased export volume of cereals, China has become a net exporter of cereals. China had a net export of cereals of 2.85 million tons in 2006, with the imported cereals being mostly used for the adjustment of grain varieties.

In terms of the correlation between international food prices and the net import volumes of different varieties of food from China, China's net import volume of food showed the same fluctuation direction with international food prices. In terms of different varieties, the net import volume of soybean has the strongest correlation with international food prices with a correlation coefficient of 0.5 (soybean is included in the category of grain and oil rather than the category of grain), followed by rice with a correlation coefficient of 0.4, wheat with a correlation coefficient of 0.3 and corn with a correlation coefficient of 0.2. It can be seen that China's net import of grains exerts certain positive impacts on international food prices. Although the increase in China's net import of food will promote the rise in global food prices, the impact is rather small and such prices will not be greatly affected, which can be proved by relevant figures. Besides, as China's import volume is relatively limited, its impact on global food prices is also limited.

In normal conditions, China's food self-sufficient rate will not be lower than 95% so as to guarantee the national food supply security. Many years of experiences have proved that reduced or rationed food supply and large food price fluctuations will occur in China when its food self-sufficiency rate falls below 95%. Therefore, China should take the food self-sufficiency rate of 95% as the cordon of China's food supply security range, and its net food import volume should be kept below 5% of the aggregate domestic food consumption. The food self-sufficient rate of 95%, which has been used by China for quite a long time, is both a food self-sufficiency rate essential for

China's food security and a promise made by China as a responsible large country.

1.2. *The framework of problem analysis*

In reality, China does face a possibility of food shortage. When such a situation occurs, China would need to solve the problem of food gap through the international market, that is, to relieve the temporary rationed/reduced food supply through large amounts of food import. Two conditions are necessary for food purchase from the international market: (a) abundant foreign exchange reserve for food imports and (b) food merchants or countries which could supply food. If China has abundant foreign exchange reserve and if there is no food embargo or trade protection (mainly export restriction), China could solve the problem of domestic food shortage through importing food from the international market. Our theoretical discussion on settlement of food gap is to prove the low probability of the occurrence of China's foreign exchange reserve shortage, food embargo and trade protection.

2. Adequate International Food Supply

2.1. *The continuous increase of world food output*

Since 1961, world's cereal production showed a continuous rising trend, increasing by 2.1 times from 0.88 billion tons to 1.82 billion tons during the rapid growth period of grain output from 1961 to 1985. The growth rate of food output globally witnessed a slight decrease since 1986 mainly because developed countries' enthusiasm in food production fell. Since the beginning of the late 20th century, though per unit area grain yield increased slightly, world's grain sown area, the total grain output and per capita food output presented a declining trend. Annual global food output increased continuously at the beginning of the 21st century and has exceeded 2 billion tons. From 1961 to 2006, the global average per unit area yield rose from 1353 kilograms per hectare to 3.5 tons per hectare, and the global average cereal per unit area yield increased by 1.6 times. The growth rate of the per unit area yield of the world now shows a declining trend.

According to relevant research and predications, the annual global grain output would maintain a growth rate of 1.8% in the following years and reach

2.34 billion tons in 2010; the growth rate of grain consumption during the same period will not exceed 1.6% and world grain consumption will reach 2.23 billion tons in 2010. With the continuous increase in grain output, the international grain trading volume will also ascend continuously. If the dependence degree of world grain trade remains the same with the present levels, world's grain export volume will reach 0.33 billion tons and world's grain import volume will reach 0.32 billion tons in 2010, thus showing a trend of grain supply outstripping grain demand. FAO also believed that no unconquerable resource and technological restrictions exist in the world as a whole. Therefore, the growth rate of global food output might exceed 1.8% in 2010, whereas the growth rate of global food consumption will gradually descend to 1.3% in 2010 and to lower levels afterwards. In general, the growth of world's food production will outstrip the increase in food consumption. Therefore, the traditional Malthusian Theory of Population has gone out of date. The present situation is not the limitation of population by food output but the inadequate demand for food. As food output is conditioned by food demand, many developed countries have started to leave part of its land fallow so as to reduce the cultivated area and agricultural input, leading to the decrease in food output accordingly. As long as there is demand for food, more food will be produced. At present, there are about 1.5 billion hectares of cultivated land in the world, merely 0.7 billion hectares of which is being actually used for food production. This indicates that there is plenty of cultivated land in the world which could be used for food production. International food production has a large potential and is completely capable of meeting the demand for food in the future.

2.2. The low correlation coefficient between the fluctuations of China's cereal output and world's cereal output

Though there is abundant food supply in the international food market in general, fluctuations do exist in international food production between different years, thus affecting food supply in international food market. The correlation coefficient between the fluctuations of cereal output of China and the world shows that their fluctuation directions was the same in merely 10 years and was opposite in most of the remaining years.

In terms of different varieties of cereals, no significant linear correlation exists between the fluctuations in the production of China's major grain crops between different years and the fluctuations of international (excluding China) grain output between different years. The correlation coefficients for the fluctuations in rice and corn output are both negative, indicating that import and export of rice and corn could be economically profitable by making use of the reverse fluctuations of their output between different years and the consequent price differences. It should be noticed that the correlation coefficients for fluctuations in rice output is as high as 0.4. However, a further analysis shows that the availability of food in major grain producing countries constitutes the main part of the food supplies in the international food market. The correlation coefficients between the fluctuations in China's corn output between different years with the fluctuations in the corn outputs of five major grain exporting countries, namely, the U.S., Argentina, etc., between different years are all negative. Although the correlation did not reach the significant negative degree of "one decreases while the other increases", the non-correlation of the fluctuations in production implies that China will be able to obtain basically normal food supplies from grain exporting countries in the international market during years of poor harvests. The correlation coefficients between the fluctuations of China's rice output with the fluctuations in the rice outputs of other major rice exporting countries in the world are all negative. The fluctuations in China's rice output between different years show certain complementarity with that of some of these countries, which is manifested by significant negative correlations. Therefore, there is no risk of supply availability in the international rice market.

The fluctuations in China's wheat output in comparison with the long-term trend have certain correlations with the fluctuations in wheat output globally, although they did not obviously go hand-in-hand with the fluctuations of world wheat output. The fluctuations in the wheat outputs of major wheat exporting countries in the world and China indicate that obvious differences exist between the fluctuations in the wheat outputs of major wheat exporting countries and the fluctuations in China's wheat output and that their correlation coefficients all stand below 0.3, belonging to the category of "non-correlation" in theory. The correlation coefficients among most of these wheat exporting countries are not high. The research shows that the correlation between the fluctuations in the wheat outputs of major wheat exporting countries in the world and the fluctuations in China's wheat output is also rather weak. Thus, it is highly unlikely for the major wheat importing countries to compete with China for food supplies in the international market in a particular year, and it's basically safe for China to ensure stable food supply in the domestic market by making use of the international wheat market.

3. Abundant Foreign Exchange Income Earned Through Exports

3.1. *China's foreign exchange income earned through exports*

In this section, we will study China's grain purchase capacity through an analysis of the scale and growth of China's annual foreign exchange income earned through exports. Generally speaking, China's foreign exchange income earned through exports increased continuously and its development could be divided into two stages. Prior to 1999, China's annual foreign exchange income increased rapidly by 3.1 times in 11 years, rising from $47.52 billion in 1988, piercing $100 billion in 1994 and gradually reaching $194.93 billion in 1999 with an annual average growth rate of 13.7%. After entering the 21st century and especially after China's Accession to the WTO, China's annual foreign exchange income through exports went up rapidly by 5.3 times in merely eight years, exceeding $200 billion in 2000, piercing $300 billion in 2002 and reaching $1218 billion in 2007 with an annual average growth rate of 25.7%.

Judging from trends during 1999–2007, China's foreign exchange income through exports will continue to rise albeit at a slightly lower growth rate. In view of China's future export situation and the pressure for RMB appreciation, based on our estimation and calculation of the mean of the data of the second stage, we predict that China's annual foreign exchange income earned through exports will reach $1233.7 billion in the short-term.

3.2. *Amount of foreign exchange required to eliminate the grain gap*

Based on the overall layout of this report and using the largest grain gap (in the case of once in one hundred year) of 57.88 million tons as the standard, we could analyze China's grain import strategies from the following two perspectives. The first perspective is that China imports one grain variety only to satisfy the demand of China's grain gap. The second perspective is that China imports grains at the proportions of major grain varieties in recent 20 years. In the first case, if China imports 57.88 million tons of rice only and if we make the calculation at the highest international market price market in recent 30 years, that is, $338 per ton in 1996, China will need $19.56 billion; if China imports 57.88 million tons of wheat only and if we make the calculation at the highest international market price in recent 30 years, that is, $207 per ton in 1995, China will need $11.98 billion; if China imports 57.88 million tons of corn only and if we make the calculation at the highest international market price in recent 30 years, that is, $164 per ton in 1995, China will need $9.49 billion. In the second case, if we make the calculation at the import proportions of wheat, rice and corn in recent 20 years, that is, 88.4%, 6% and 5.6%, their import volume would be 51.187 million tons, 3.473 million tons and 3.225 million tons respectively; if we make the calculation at their highest prices in recent 20 years, that is, $207 per ton, $338 per ton and $164 per ton, China will need $1.06 billion, $0.117 billion and $0.53 billion respectively, the sum of which is $12.3 billion. Therefore, according to conservative estimation, China will need $19.56 billion at the most to eliminate the grain gap which occurs once in 100 years.

3.3. *China has foreign exchange sources to purchase grain*

China's foreign exchange income earned through exports is as high as $1233.7 billion, far higher than $19.56 billion which is needed for China's grain import. Besides, China's foreign exchange reserve accelerated its growth after the foreign exchange management system reform in 1994. Traditional views believe that the foreign exchange reserve of a country should not be lower than the country's total import payment in foreign exchange of three to four months. After the Asian Financial Crisis, it was proposed that the foreign exchange reserve of a country should not be lower than the country's short-term external debts. In accordance with these opinions, China's foreign exchange reserve should not be less than

$200 billion. By the end of June 2008, with a foreign exchange reserve of $1808.8 billion, China's foreign exchange reserve had changed from "shortage" to "surplus". Obviously, China has abundant foreign exchange reserve.[3] Therefore, both in terms of the total amount of foreign exchange reserve and the total amount of foreign exchange income earned through exports, China has enough foreign exchange to purchase grains, and the situation of China having nil or insufficient foreign exchange to purchase grains will not occur at all.

4. The Possiblity of Unblocked Grain Import Channels

4.1. *An analysis of food embargos*

4.1.1. *Food embargos in the world*

Grain embargo generally refers to the reduction in the amount of or even discontinuation of grain circulation in the international market caused by policy interventions. In the history of world food trade, large grain exporting countries reduced or even discontinued their grain supply to grain importing countries for the sake of their political goals, sometimes solely or mainly in the form of grain trade and sometimes in the form of a package of embargos. Grain embargo has four aims: to "punish" or "revenge on" the grain importing countries, to safeguard the economic interests and political stability of grain exporting countries, to interfere in the internal affairs of and affect the political situation of grain importing countries; to force the recipient countries to adjust their policies in accordance with the wishes of grain exporting countries through restricting or stopping the so-called "concession export" of grain which took on the nature of aids to these countries (Hu Yingchun, 2003).

---

[3]Based on China's foreign exchange reserve data from 1980 to 2007, we established an empirical normal distribution function in accordance with different intervals of China's foreign exchange reserve (classification by quantity segments) to predict China's foreign exchange reserve in the following twenty years (2009–2028). We used the mean as the testing data and proved that the probability of the occurrence of insufficient foreign exchange reserve was 0.954%.

The research of Hufbauer and Schott (1985) suggested that 103 cases of economic sanctions occurred throughout the world from 1919 to 1984. Among the 84 cases of economic sanctions occurred from 1950 to 1985, 10 cases involved food embargos. Among the 10 cases of food embargos, nine were developing countries and the remaining one was the Soviet Union; eight cases were initiated by the U.S. (Winters, 1990) and one case was China's imposition of food embargo on Vietnam during the Sino-Vietnamese War; five food embargos totally failed, two caused short-term pressure on embargo targets without any long-term influence, one succeeded with negligible effect, and two basically succeeded (Winters, 1990, p. 176).

At least half of the 10 food embargos belonged to partial or complete discontinuation of food export for aid purposes rather than commercial embargos on food export. The remaining are mostly food embargos in the broad sense, that is, embargos targeted trade of large numbers of all the commodities including food. Only one belongs to food embargo in the narrow sense as defined above. The two successful grain embargos have a common feature, that is, their targets both belong to food export for aid purposes. The reasons for the failure of grain embargos include the interchangeability and substitutability of grains, interest divergence among grain exporting countries, humanitarian considerations, the fact that grain embargo is a double-edged sword in that grain embargo on a country in the WTO era implies the economic loss of major grain producing countries.

4.1.2. *Analysis of the possibility of food embargo on China*

After China's entry into the WTO, China's food foreign trade dependence degree continued to rise. After a comprehensive investigation of the reasons and effects of food embargo as well as the national conditions of China, it can be seen that the possibility of food embargo on China does exist, but this possibility is negligible. A number of scholars believe that food is usually excluded from the embargo list in the past for humanitarian reasons, and the probability of occurrence of food embargo is not high under current social conditions. However, it is possible for the reduction of food supply caused by economic sanctions to happen. History suggests that China is very unlikely to face food embargos in the future.

4.1.3. *Strategies to solve the problem of food embargos*

Regarding the decrease in food supply caused by economic sanctions, if the economic sanctions are carried out jointly by several countries, the availability of foreign resources for China will be reduced, but China will not be greatly affected in terms of food supply. Instead, countries which carry out food embargos will suffer immense losses in foreign trade. For instance, food sanctions imposed on the Soviet Union by the U.S. caused the U.S. to suffer a loss of $2.2 billion. When China's food self-sufficiency rate remains above 95%, it could easily acquire food supply from other countries in the world. If China's food self-sufficiency rate stands above 90%, apart from international trade, China could also make use of food stocks to maintain a food supply for two months; during this period, the rising food price will increase investment in grain planting in China, including input of human resources and capital, etc., thus guaranteeing the increase in food supply within a short period.

The imposition of joint blockade centered around the United Nations on China will pose certain challenges to China's food supply, but China still has the ability to dissolve it. Economic sanctions will lead to very different results in China from that of Iraq. Since Iraq depended heavily on food imports, economic sanctions led to severe social unrest in Iraq. The UN's sanctions on Iran and North Korea did not include food embargos and allowed the supply of foodstuffs in humanitarian circumstances which were subject to authorization by the UNSC. The case of Iraq will not happen to China because China's food self-sufficiency rate generally will not fall below 90% and would most likely remain above 95%. Under current food trade conditions, the normally functioning granaries will be able to offer food supplies for two months and fill the short-term urgent needs. Besides, as China does not rely on import for cereals and its major import grain variety which has the largest annual import volume at present is soybean which is mainly used to produce edible oil, China's cereals supply will not be greatly affected if food embargo occurs and it won't be a problem to prevent its people from starvation. Regarding soybean, on one hand, China could meet the demand by using stocks; on the other hand, China could promote the increased investment in soybean planting and improve the supply of soybean within a quarter.

Therefore, even if food embargos on China really happen, China's self-sufficiency rate will be able to stand the test and prevent the occurrence of food crisis in China. If the food self-sufficiency rate stands above 95%, China would not be affected greatly; if it falls between 90% and 95%, China will be able to get through the difficulty within a short period of time; if it falls below 90%, China will pay a heavy price to cope with the crisis. However, no substantial threat exists. Therefore, if the possibility of the imposition of food embargo on China is high, China has to ensure that its food sufficient rate stands above 95% so as to guarantee China's domestic stability and its fulfillment of the responsibility of a large country. The fact, however, is that the possibility of the imposition of food embargo on China is negligible.

4.2. *An analysis of the existence of trade protection*[4]

4.2.1. *Summary of export protection*

In an essence, the restriction on food export is the limitation of external demand through government intervention. Under the condition of unchanged domestic good output, the restriction on food export will ensure that domestic food output is mainly used to satisfy domestic food demand and bring down food prices. This sort of logic emerged several times in China's food system reform. When food prices went up, the government forced down food prices through administrative interventions and farmers' enthusiasm for production was dampened; the government then tried to liberalize the food market and could not help to start another round of administrative interventions when the market-led food prices rose again. This is what Lu Feng called "the Semi-cycle Reform Phenomenon". The restriction on food export could be seen as a derivative of this sort of

---

[4]We hypothesize the probability of the occurrence of domestic trade protection conforms to binomial distribution. We assume that the international and domestic food prices will both go up when neither the international food output nor China's food output could satisfy people's need for production and living due to the negative growth rates of food output of both China and the world during the same period. Based on the data from 1962 to 2006, we calculated that the probability of domestic food trade protection is very low, standing at merely 1.41382E-30, almost equivalent to zero.

actions. The forceful elimination of external demand by the government will bring down domestic food prices in the short-term. However, as the American economist Schultz put it, "All farmers are the best economists". The rational decision of farmers is: in the next production cycle, if the cost constraint remains unchanged and the expected market price is low, the expected profit will decline and losses may even occur; therefore, the input in food production will definitely be reduced; this way, food supply in the next production cycle will be insufficient, and it will be highly possible for food prices to pick up. Therefore, the prohibition of food export will promote further increase in international prices through supply reduction in the short-term and production repression in the medium and long-terms.

The restriction in food exports will also reduce the income of domestic farmers and go against the increased welfare measures for producers. Moreover, the original share of the international market will be surrendered to other countries, and the international competitiveness will inevitably be weakened. The governmental restriction in food export and prohibition in the participation of agriculture in international competition will definitely cause the recession of agriculture in the long run and low efficiency in utilization of resources. Under the layout of global economic integration, the low efficiency of utilization of domestic agricultural resources and the high efficiency in the international market cannot ensure that countries restricting food exports would smoothly avoid the strike of possible food insecurity.

Restriction in food exports violates the free trade spirit of the WTO. Moreover, the food trade policies of some countries also violate the prevision of Article 12 of the six part of the US–China WTO Agreement on Agriculture. In accordance with paragraph 2 of Article XI of GATT 1994, where any member institutes any new export prohibition or restriction on foodstuffs, the member shall observe the following provisions: (a) the member instituting the export prohibition or restriction shall give due consideration to the effects of such prohibition or restriction on importing members' food security; (b) before any member institutes an export prohibition or restriction, it shall give notice in writing, as far in advance as practicable, to the Committee on Agriculture comprising such information as the nature and the duration of such measure, and shall consult, upon request, with any other member having a substantial interest as an importer with respect to any matter related to the measure in question. The member instituting such export prohibition or restriction shall provide, upon request, such a member with necessary information. The provisions of this Article shall not apply to any developing country member, unless the measure is taken by a developing country member which is a net-food exporter of the specific foodstuff concerned.

Therefore, policies and measures restricting food exports are not effective ways to solve the problem of food security. Moreover, they violate relevant rules of the WTO. The fundamental means to guarantee international food supply include the active promotion of international agricultural cooperation, the integration of international resources, the increase of agricultural investment, the enhancement of agricultural infrastructure and the increase of agricultural scientific input.

4.2.2. *Explanations for recent food trade protection*

The continuously rising international food price since 2006 led to the worldwide limitations on food export and trade protection in the beginning of 2008. Part of the major grain producing countries cut down their grain export volumes substantially. In the beginning of 2007, India, the second largest wheat and rice producing country, declared the prohibition of wheat export during 2007 and then the prohibition of corn export. In October 2007, it prohibited export of flour. Afterwards, Argentina, a large exporting country of agricultural products, and Russia, the fifth largest exporting country of wheat in the world, also declared to increase their export tax on partial grain products. China also joined the group of grain export restricting countries at the end of 2007. In March 2008, Vietnam, India, Cambodia, Egypt and other major rice exporting countries issued policies on restriction of rice export in quick succession (Li Zhe, 2008).

In terms of food output, China's food output went up continuously, but the growth rate went down slightly; world's food output also ascended gradually, but the increase was limited. Generally speaking, although world's food output continued to increase, the grain planting structure has changed. In particular, the U.S., a large grain producing country, started to plant large amounts of corn instead of soybean. China has also started to develop corn-based biofuel production industry in the recent two years. Corn is rarely used directly as food for human beings and cannot be easily converted to food when edible food shortage occurs. In addition, the constraint of growth cycle exists in food production. Therefore, the excessive development of bioenergy led to the rising international food price and the insufficient market supply of food.

4.3. *The unblocked trade*

With the guidance of market rules, the relationship between food supply and demand will lead in the smooth development of food trade. When food gap occurs in China and if there is no food embargo, China could make use of its abundant foreign exchange reserve (which comes from foreign exchange incomes) to purchase the required food from the international market and solve the problem of possible food shortage in China. Even if the international community does impose food embargo on China, international humanitarian aid will be provided to China to solve part of the problem of food shortage and China has ample food reserve to cope with short-term difficulties. Moreover, the possibility of food embargo is negligible. Therefore, generally speaking, China could smoothly purchase necessary food from the international market when food gap occurs in China.

5. The Feasibility of Settling the Problem of Food Gap through International Trade

The settlement of China's gap in food supply through international means is generally feasible. First, the increase in aggregate global food output is assured. The local food crisis also reminds countries to ensure their food supply, and these countries undoubtedly will continue to increase food production. Second, the correlation between the fluctuations of food output of China and the world is rather weak, indicating that food supply of other countries will remain sufficient even when food shortage occurs in China. Moreover, China has abundant foreign exchange reserve; the possibility of the imposition of food embargo on China by the international community is rather low; it is impossible for domestic trade protection to happen throughout the world. Therefore, China could make use of its abundant foreign exchange reserve to purchase food from the international food market. The international market will be able to solve the problems of China's food gap. No substantial obstacle exists in this aspect.

Conclusion

China is a country with a large population. We cannot emphasize how important it is to supply ample food to our people and guarantee food security. In particular, the Great Famines in the 1950s and 1960s have left an indelible imprint upon the memory of Chinese people. Even the current highest level of decision-making in China cannot forget it. China's institutional reform originated from the vast countryside also because food shortage constituted an urgent social problem to be solved at that time. However, the significance of food does not equal to the significant of food security. After 30 years, great changes have happened to domestic and international environment, and the importance of food security has increasingly taken a back seat among a large number of major national policies.

It is definitely wrong and extremely harmful to equal food security to cultivated land protection or to attribute cultivated land protection mostly to food security because this opinion will hinder the exploration and recognition of the real reasons. This report fully approves of cultivated land protection if it is limited to maintenance and economical and efficient utilization of cultivated land. Nevertheless, in present public opinion and policies, cultivated land protection is obviously confined as the strict control of the amount of cultivated land so that 1.8 billion mu was even defined as "an impassable red line". Cultivated land protection in this sense and the direct association of cultivated land protection with food security are negated and criticized by this report.

In the long term, the global food supply and demand will be basically balanced. In the following years, food supply may even slightly outstrip food demand. It was estimated that the annual world food output would maintain a growth rate of 1.8% in the following years and reach 2.34 billion tons by 2010 and that the growth rate of the annual world food consumption would not exceed 1.6% during the same period and reach 2.225 billion tons by 2010. More importantly, the annual world cereal sown area has stabilized at around 0.7 billion tons since the 1960s, accounting for about 50% of the total cultivated land area of the world. Despite the rapid expansion of world's population, the global per capita grain output has increased to a certain extent, rising from 310 kilograms in the 1960s to 339 kilograms in 2006 and leading to the decline of relative world's grain prices in the long run. This indicates that we are getting rid of the constraint of Malthus' law of population thanks to the advancement of agricultural science and technology. This conclusion lays a basis for the analysis of this report.

Many factors may pose a major threat to food security, such as, wars, internal strife, plague, and so on. However, these factors should be deducted from the logical analysis of the relationship between cultivated land protection and food security because the total amount of cultivated land obviously will not affect the total grain output once the abovementioned *force majeure* events occur. Surely, the significance of economic system for food security could not be underestimated. It should be noted that the Great Famine in the 1950s and the 1960s occurred under China's planned economic system at that time. Nevertheless, since China's reform and opening-up has progressed to the present status, there's no reason to believe that fundamental reversal might happen to China's economic system. Therefore, this report actually discussed the relationship between the quantity and quality of cultivated land with grain output under ordinary conditions.

Under the abovementioned premise, this report defines the national food security as a state's capability of closing the gap between food supply and demand. In this case, we need to measure and calculate the gap or the gap rate between food demand and supply. Based on the analysis of the relevant empirical data of China, we calculated the food gap rates (the ratio of food gap to food output) when food shortage occurs once in 10 years, 50 years and 100 years. In accordance with different calibers, the food gap rates are 5.60%, 9.99% and 11.59% and 6.72%, 10.89% and 12.41% respectively. This calculation result is apparently lower than the general subjective surmise.

We can solve the abovementioned gap between food supply and demand through both domestic and international means. Domestic means include the handling of food stock and the increase of agricultural factor input. International means refer to foreign trade, namely, food import. In reality, China's current food stock stands between 0.15 and 0.2 million tons, making up 30%–40% of China's annual food output. Therefore, we will be fully capable of coping with food crisis which occurs once in 100 years merely through changes in domestic food stock.

The core meaning of market economy is "substitution", that is, the substitution among different factor inputs under a given output and the substitution among the consumption of different products under a given utility. Apparently, as one of the agricultural factors, cultivated land has certain scarcity but could be substituted by other factor inputs to a certain extent. After measurement and calculation, we conclude that even when food crisis which happens once in 100 years occurs and when grain sown area reduces by 10%, China can conquer this crisis by increasing the inputs of other factors by 15% (in accordance with different calibers, this proportion is 14.4% and 15.5% respectively).

On the basis of domestic measures, food crisis will be more easily solved by importing food through foreign trade. With the food prices spiraling in recent 30 years as the base price, when food crisis which happens once in 100 years occurs, China could eliminate its food gap by spending $19.56 billion on food import. As China's annual foreign exchange income stands at $1200 billion excess, this expenditure will not constitute any substantial problem in comparison with the rapidly developing Chinese economy. Surely, we should also take into account the pressure of the rising international food prices caused by China's large amounts of food import. This problem could be solved by supplementing China's food stock through food import in batches. Besides, in the long term, the rising trend of international food prices will inevitably lead to the increase in global (including domestic) food supply because the global food output and the global food trading volume are mainly limited by demand constraints rather than resource constraints.

Under the complicated and volatile international situation, it is reasonable for China to be fully alert to the possibility of the imposition of food embargo measures on China by the international community.

Nevertheless, experimental facts show that among the 84 cases of economic sanctions occurred from 1950 to 1985, merely 10 cases involved food embargos, accounting for less than 12% of the total number of economic sanctions; at least half of the 10 food embargos belonged to partial or complete discontinuation of food export for aid purposes rather than commercial embargos on food export. As food embargos obviously violate the international principles of humanism and the effects of food embargos are limited, it could be estimated that the occurrence of food embargos will be further reduced in the future. Therefore, it is highly impossible for China, a country with strong negotiation and game capability, to encounter the sanction of food embargo imposed by the international community; but the possibility of food supply reduction in China caused by other sanctions still exist to a certain extent.

Recently, many new examples of countries setting up food trade barriers towards one another (joint limitation on food export) have emerged. This is mainly because of the exploitation of biofuels, the enhanced environment protection measures, devaluation of U.S. dollars and the vicious transactions of food futures. However, facts have proved that actions violating the principle of mutual benefit and free trade are not sustainable. With the plunge in international food future prices, the farce of countries setting up food trade barriers towards one another has come to an end. After the event, we reflect that this was merely a false alarm.

In general, under the mechanism of free trade and substitution of factors of market economy and the condition that China has sufficient food stock and foreign exchange income, food security problems basically will not happen to China. Food insecurity or great famine only occurs in China during the period of planned economy when Chinese economy was not liberalized. It is groundless to ignore the substitutability of cultivated land as an agricultural factor and to associate cultivated land protection directly with food security both in theoretical and experimental terms. The red line of cultivated land area defined by the government will definitely be broken through. In addition, the previous red line of 2.1 billion mu has already been broken through because what it constrained was merely the industrialization and urbanization process of China.

Obviously, due to their monopoly of the primary land market, the people's governments at all levels (especially local governments) made immense profits through collection of land transfer fees and cultivated land occupation tax and low compensation for land-expropriated farmers, thus greatly stimulating their enthusiasm of "city management" for the rolling development of rural land. Large amounts of cultivated land was actually occupied for other purposes or effectively utilized during this process. Therefore, the central government's initial intention of attempting to prevent the occurrence of the above phenomenon through setting up the red line of cultivated land area deserves corresponding respect. Nevertheless, it is rather ridiculous to equal cultivated land protection to food security or attribute cultivated land protection to food security. In fact, the emphasis of cultivated land protection policies should be placed on the protection of farmers' employment rights which is embodied in land rather than food security.

References

Alan de Brauw, Jikun Huang and Scott Rozelle (2000). Responsiveness, flexibility, and market liberalization in China's agriculture. *American Journal of Agricultural Economics*, 82(5), 1133–1139.

Alex F. McCalla (1982). Food Security for Development Countries. By Alberto Valdes. *Journal of Economics Literature*, 20(2), 622–624.

Amartya Sen. The food problem: Theory and policy. *Third World Quarterly*, 4(3), 447–459.

C. Peter Timmer. Food Security and Economic Growth: An Asian perspective.

C. Peter Timmer (2000). The macro dimensions of food security: Economic growth, equitable distribution, and food price stability. *Food Policy*, 25(3), 283–295.

Carlton G. Davis, Clive Y. Thomas and William A. Amponsah (2001). Globalization and poverty: Lessons from the theory and practice of food security. *American Journal of Agricultural Economics*, 83(3), 714–721.

Frank Riely, Nancy Mock, Bruce Cogill, Laura Bailey and Eric Kenefick (1999). *Food Security Indicators and Framework for Use in the Monitoring and Evaluation of Food Aid Programs*. USA: USAID.

Handy Williamson, Jr (2001). Globalization and poverty: Lessons from the theory and practice of food security: Discussion. *American Journal of Agricultural Economics*, 83(3), 730–732.

Heinz W. Arndt (2004). Memorial Lecture, Canberra, November 22.

Helen H. Jensen. Food insecurity and the food stamp program. *American Journal of Agricultural Economics*, 84(5), 1215–1228.

J. Craig Jenkins and Stephen J. Scanlan (2001). Food security in less developed countries, 1970 to 1990. *American Sociological Review*, 66(5), 718–744.

Jikun Huanga and Howarth Bouis (2001). Structural changes in the demand for food in Asia: Empirical evidence from Taiwan. *Agricultural Economics*, 26, 57–69.

Joachim von Braun (2007). The world food situation: New driving forces and required actions. A report of annual general meeting of the Consultative Group on International Agricultural Research (CGIAR).

Joachim von Braun, M.S. Swaminathan and Mark W. Rosegrant. The agriculture, food security, nutrition and the millennium development goals. IFPRI Annual Report 2003–2004.

Joachim von Braun (1990). Food insecurity: Discussion. *American Journal of Agricultural Economics*, 72(5), 1323–1324.

Lin, J.Y. and Denis Yang (2000). *Food Availability, Entitlements and Chinese Famine of 1959–1961*.

Lin J.Y. (1992). Rural Reform and Agriculture Growth in China. *American Economic Review*, 82, 34–51.

Lin, J.Y. (1992). *Institution, Technology and Agricultural Development in China*. The Joint Publishing Co. Ltd., 1992.

Lisa C. Smith, Amani E. El Obeid and Helen H. Jensen (2000). The geography and causes of food insecurity in developing countries. *Agricultural Economics*, 22(2), 199–215.

Luther Tweeten (1999). The economics of global food security. *Review of Agricultural Economics*, 21(2), 473–448.

Randolph Barker (1982). *Food security: Theory, policy, and perspectives from Asia and the Pacific Rim*, Anthony H. Chisholm and Rodney Tyers (eds.), p. 359. Lexington, MA: Lexington Books.

Randolph Barker (1984). Food security: Theory, policy, and perspectives from Asia and the Pacific Rim. *Journal of Development Economics*, 14(1), 273–277.

Reutlinger, S. and Selowsky, M. (1976). Malnutrition and poverty: Magnitude and policy options. World Bank Occasional Paper No. 23, Johns Hopkins University Press for the World Bank, Baltimore, MD.

Robert Paarlberg (2000). The weak link between world food markets and world food security. *Food Policy*, 25(3), 317–335.

Robert W. Herdt (1984). Differing perspectives on the world food problem: Discussion. *American Journal of Agricultural Economics*, 66(2), 186–187.

Songqing Jin, Jikun Huang, Ruifa Hu and Scott Rozelle (2002). The creation and spread of technology and total factor productivity in China's agriculture. *American Journal of Agricultural Economics*, 84(4), 916–930.

Sonya Kostova Huffman and Helen H. Jensen (2008). Food assistance programs and outcomes in the context of welfare reform. *Social Science Quarterly*, 89(1), 95–115.

Wen, G.Z. (1993). Total factor productivity change in China farming sector: 1952–1989. *Economic Development and Cultural Change*, 42, 1–41.

Winters, L.A. (1990). Digging for victory: Agricultural policy and national security. *The World Economy*, 13, 170–190.

You, Zhikang, James E. Epperson and Chung L. Huang (1998). Consumer demand for fresh fruit and vegetables in the united stipulates. *Geogia Agriculture Experiment Station Bulletin*, 431.

Beijing Flower Industry Striding Towards Urban Agriculture (2007). China Flower & Gardening News, 2 June (in Chinese).

Beijing municipal people's procuratorate carried out thorough inspection of three fields with serious corruption: Health care, land and traffic (9 October 2004). *Legal Evening News* (in Chinese).

Sheng Hong (2006). How much should be paid to land-expropriated farmers as compensation? *China Review*, March (in Chinese).

Assessment of World Food Security Situation, Report of the 33 Session of the Committee on World Food Security, 7–10 May 2007, Rome.

Loren Brandt, Li Guo, Huang Jikun and Scott Rozelle (2004). Land tenure and transfer rights in China: An assessment of the issues. *The Quarterly Journal of Economics*, 3(4), 951–983 (in Chinese).

Beijing Municipal Bureau of Statistics (2006). *Beijing Statistical Yearbook 2006*. Available at http://www.bjstats.gov.cn/tjnj/2006-tjnj/ (in Chinese).

Bi Yuyun and Zheng Zhenyuan (2000). An analysis of the changes of the real area of China's cultivated land since the founding of the New China. *Resources Science*, 22(2) (in Chinese).

Cao Shuji (2005). The death toll of China between 1959 and 1961 and the causes. *Chinese Journal of Population Science*, 1 (in Chinese) 14–28.

Chen Jiansheng (2006). *Conversion of Cultivated Land and the Sustainable Development of the West*. Chengdu: Southwestern University of Finance and Economics Press (in Chinese).

The Office of Vegetable of the People's Government of Daxing District (2006). Excellent arrangement of crops and significant benefits. *Daxing Vegetable*, 8 (in Chinese).

Deng Xiangzheng, Huang Jikun and Scott Rozelle (2005). Cultivated Land Conversion and Bio-productivity in China-An Discussion of China's Food Security. China Soft Science Magazine, 5 (in Chinese).

Deng Yiwu (2004). *Research on Macro Control and Regulation of Grain*. Beijing: Economy and Management Publishing House (in Chinese).

Dong Mei and Zhong Funing. An empirical study of the economic sustainability of the policy of converting cultivated land to forests — Taking Ningxia Hui autonomous region as an example. Paper prepared for the 4th Chinese Economics Annual Conference (in Chinese).

Fan Shenggen and M.A. Sombilla (1997). China's food supply and demand in the 21st century: Baseline projections and policy simulations. Paper prepared for the American Agricultural Economics Association Annual Meeting, July 31.

Fan Ziying and Meng Lingjie (2005). A review of researches on the great famine from 1959 to 1961 in China. *China Rural Survey*, 1 (in Chinese).

Feng Zhiming, Liu Baoqin and Yang Yanzhao. A study of the changing trend of China's cultivated land amount and data reconstructing: 1949–2003. *Journal of Natural Resources*, 20(1) (in Chinese).

Fu Chao, Zheng Juan'er and Wu Cifang (2007). Historical investigation and revelation of quantity change of cultivated land resources in China since 1949. *Scientific and Technological Management of Land and Resources*, 2007 (in Chinese).

Gale Johnson (1999). East Asia. Is there sufficient food supply? *Chinese Rural Economy*, 12 (in Chinese).

Gao Jikai and Puwei *et al.* (2007). The blooming flowers make both urban and rural areas fragrant. *Chengdu Daily*, 16 May (in Chinese).

Rural Economic and Social Survey Corps of the National Bureau of Statistics of China (2003). *Research on China's Food Problems*. Beijing: China Statistical Publishing House (in Chinese).

Hu Yingchun and Liu Qing (2003). The risks of food embargo and the policy choice of China's food security. *China State Farms Economy*, 8 (in Chinese).

Huang Jikun (2004). China's food security 2004. *Chinese Rural Economy*, 4–10 (in Chinese).

Huang Jikun *et al.* (2006). *China's Agricultural and Rural Development in the 21st Century.* Bijing: China Agriculture Press (in Chinese).

Huang Jikun and Luo Zeer (1998). *China's Grain Economy Entering the 21st Century.* Beijing: China Agriculture Press (in Chinese).

Jiang Boying (2004). *Deng Zihui and China's Rural Reform.* Fujian: Fujian People's Publishing House (in Chinese).

Jin Hui (1998). Three years of favorable weather-a meteorological and hydrographic investigation of the period from 1959 to 1961. *Methods*, 3 (in Chinese).

Jin Songqing and Klaus Deininger (2004). The development of China's rural land leasing market and its implications on fairness and efficiency of land use. *Quarterly Journal of Economics*, 3(4), 1003–1027 (in Chinese).

Li Jinping (2006). The Truth of the Three Years of "Natural Disasters". Available at http://www.ifeng.com/phoenixtv/83931275940855808/20060505/790687.shtml.

Li Qiang and Zhang Haihui (2004). China's urban layout and the society of high population density. *Strategy and Management*, 2004 (in Chinese).

Li Ruojian (1999). The Climax and Decline of Urbanization during the Great Leap Forward. *Population and Economy*, 5 (in Chinese).

Li Ruojian (2001). China's grain output, consumption and distribution during the great leap forward and the hard times. *Journal of Sun Yat-sen University* (Social Science Edition), 6 (in Chinese).

Li Shidong (2004). *Research on Conversion of Cultivated Land to Forests in China.* Science Press (in Chinese).

Li Siheng (2003). World food situation in 2003 and an analysis of the two years after China's entry into the WTO. *China Grain Economy*, 12 (in Chinese).

Li Zhe (2008). The itensified imbalance of world food supply and demand. *Securities Times*, 2nd April (in Chinese).

Li Weimin, Li Shuying, *et al.* (1988). *An Introduction to World Food Security.* Beijing: China Renmin University Press (in Chinese).

Liu Lifeng *et al.* (2006). Review of the development and features of China's grain import and export in recent years. *Grain Technology and Economy*, 4 (in Chinese).

Lu Feng (2002). *Relaxing when Economic Situation is Easy and Tightening When Economic Situation Is Tough: Research on the Periodical Repetition of China's Grain and Cotton Distribution Reform, Case Studies in China's Institutional Change.* Beijing: China Financial and Economic Publishing House, Vol. 3, pp. 299–341 (in Chinese).

Lu Feng (2004). *Semi-cycle Reform Phenomenon: Research on China's Grain and Cotton Distribution Reform and Food Security.* Beijing: Peking University Press (in Chinese).

Lu Feng (1998). An evaluation of adjustments to the grain trade policy and the risks involved in grain embargo. *Social Sciences in China*, 2 (in Chinese).

Lu Feng and Xieya. The Supply of and the Demand for Grains in China and the Trend of Prices (1980~2007) — A Study on Grain Price Fluctuations. Macro Stability and Grain Security, CCER Working Paper Series, Peking University (in Chinese).

Lu Dadao. The "rush advance" of urbanization is the source of endless troubles. Available at http://www.cas.ac.cn/html/Dir/2006/12/30/14/65/68.htm [accessed on 31 December 2006].

Luo Pinghan (2006). Several problems concerning grain production and sales in the years from 1958 to 1962. *Journal of Chinese Communist Party History Studies*, 1 (in Chinese).

Mao Yushi. How to See the Great Famine of China? From *China-Review.com*. Available at http://www.china-review.com/sao.asp?id=19511 (in Chinese).

Nie Zhenbang (ed.) (2005). *Training Tutorial in the Regulations on the Administration of Grain Distribution*. Beijing: China Logistics Publishing House (in Chinese).

Nie Zhenbang (ed.). *China Grain Development Report over the Years*. Beijing: Economic & Management Publishing House (in Chinese).

The Office of Soft Science Committee of the Ministry of Agriculture of China (2001). *Grain Security Problems*. Beijing: China Agriculture Press (in Chinese).

Qin Yazhou and Wu Yong (2007). The overspeed of China's urbanization should be rectified by legislation of the central government. *China Comment* (in Chinese).

Qu Shang (2004). Review and analysis of China's food import and export since the mid-to-late 1990s. *China Grain Economy*, 2 (in Chinese).

Qu Shang (2006). A Historical Analysis of China's Grain international Trade and Its Characters. *Researches in Chinese Economic History*, 3 (in Chinese).

Editorial Office of the Grain Work in Contemporary China of the Ministry of Commerce of China (1989). *Historic Materials for Grain Work in Contemporary China*, 152, 156, 315. Beijing: China Agriculture Press (in Chinese).

Shang Qiangmin (2005). An analysis of changes of China's wheat import and export in 2004. *China Grain Economy* 1 (in Chinese).

Shao Xiaomei and Xie Junqi (2007). An analysis of the regional changing trend of China's cultivated land resources. *Resources Science* (in Chinese).

Shen Weishou, Cao Xuezhang and Shen Fayun (2006). Classification and grading of land degradation in China. *Journal of Ecology and Rural Environment*, 22(4) (in Chinese).

Sheng Hong (1992). *Division of Labor and the Transactions*. Shanghai: Shanghai Sanlian Book Store (in Chinese).

Sun Han and Shi Yulin (ed.) (2003). Agricultural Land Use in China. Nanjing: Jiangsu Science and Technology Press, 2003 (in Chinese).

Tan Jingrong (2003). Conversion of cultivated land to forests and China's food security problem. *Grain Issues Research*, 1, 6–8 (in Chinese).

Tang Jian, *et al.* (2006). *China's Cultivated Land Protection System and Policy Research*. Beijing: China Social Science Press (in Chinese).

Tang Zhengmang (2007). Chen Yun and grain import during the economic difficulties. *Extensive Collection of the Party History*, 12 (in Chinese).

Tang Zhengmang (2007). *The Seven Thousand Cadres Conference and Grain Problem*. Beijing: Corpus of Party History (in Chinese).

USAID Policy Determination, 1992. United States Department of Agriculture, 1996, P. 2.

Wang Weiluo (2001). Tian Wen: Chinese Famine of 1959–1961. *Modern China Studies*.

Wang Hongguang (2005). *The Research on China's Food Security*. Beijing: China Agriculture Press (in Chinese).

Xiao Guoan (2005). *The Research on China's Food Security*. Beijing: China Economic Publishing House (in Chinese).

Xinhuanet.com, The Population Density in Rural Areas of China Has Amounted to 847 People Per Square Kilometer (16 June 2004).

Xue Li and Wu Mingwei (2006). A discussion about the heteromorphism of the rural settlements and its policy in Jiangsu province. *Urban Planning Forum* (in Chinese).

Xue Zhiwei (2006). Watching Out Four Phenomena of Land Wastage. *Economic Daily News*, 28 June.

Ya Zhou (2003). An analysis of the influencing factors of domestic rice import. *Reclaiming and Rice Cultivation* 1 (in Chinese).

Ye Xingqing (1999). On the relationship between grain supply and demand and the regulation. *Economic Research Journal*, 8 (in Chinese).

Yu Jianrong (2006). Social conflicts in transitional China. Website of the Unirule Institute of Economics. Available at http://www.unirule.org.cn/secondweb/DWSummary.asp?DWID=14.

Yu Jingzhong (2005). Differentiated strategies of grain import and export. *Bulletin of Agricultural Science and Technology*, 10 (in Chinese).

Yu Wen and Huang Jikun (1998). Viewing China's Grain Market Reform from the Integration of Rice Market. *Economic Research Journal*, 3, 50–57 (in Chinese).

Trailing the Huge Unidentified Property of over 100 Million Yuan of Yin Guoyuan, Former Deputy Director of Shanghai Municipal Housing, Land and Resource Administration Bureau, *China Real Estate Business*, 16 April 2007 (in Chinese).

Zhang Shigong and Wang Ligang (2005). A brief analysis of the changes of China's cultivated land area and the reasons. *Chinese Journal of Agricultural Resources and Regional Planning*, (in Chinese).

Zhao Qiguo, Zhou Shenglu, Wu Shaohua, *et al.* (2006). Changes and the Sustainable Utilization of China's Cultivated Land Resources and the Protection Countermeasures. *Acta Pedol Sin* (in Chinese).

Editor Board of Almanac of China's Finance and Banking. *Almanac of China's Finance and Banking*. Beijing: China Financial Publishing House (in Chinese).

Rural Development Institute of Chinese Academy of Social Sciences, Rural Social Economy Survey Organization of National Bureau of Statistics of China (2008). *Rural Economy of China Analysis and Forecast (2007–2008)*. Beijing: Social Science Academic Press (in Chinese).

Ministry of Agriculture of the People's Republic of China (2004). China Grain Security Report, May (in Chinese).

Zhong Puning, Zhujing, *et al.* (2004). *Grain Market Reform and Globalization: Another Choice of China's Food Security*. Beijing: China Agriculture Press (in Chinese).

Zhongwen (ed.) (2005). *Bainian Chen Yun*. Beijing: Central Party Literature Press (in Chinese).

Zhou Zhangyue and Wan Guanghua (1999). A discussion of the research methodology of market integration — concurrently comment on viewing China's grain market reform from the integration of rice market by Yu Wen and Huang Jikun. *Economic Research Journal*, 3, 73–79 (in Chinese).

Zhu Jing (2003). Agricultural public investment, competitiveness and food security. *Economic Research Journal*, 1, 13–20 (in Chinese).

Zhu Xigang (2004). An analysis of the balance of food supply and demand in China. *Issues in Agricultural Economy*, 12 (in Chinese).

Zhu Xigang (1997). *Cross-Century Exploration: Research on China's Grain Issues*. Beijing: China Agriculture Press (in Chinese).

Index

The book introduces the development of regulatory farmland protection, food security policies and institutional drawbacks in a comprehensive and systematic way from historical and realistic perspectives. In addition, it offers some innovative and insightful ideas. From the perspectives of the global market, the book argues that market and privatization (reconstruction of the land ownership of farmers) are the best way to ensure food security in China.

Professor Zhang Zhenqiang
Hubei University of China

Farmland in China is a scarce resource, which needs protection, the more the better. However, the protection incurs great cost. On the one hand, farmland is protected at the expense of other land uses; on the other hand, the implementation costs of different protection methods are not the same. This book is a must read for insights into this problem.

Professor Zhou Qiren
Peking University